Adrian E. Scheidegger

Morphotectonics

Springer

Berlin
Heidelberg
New York
Hong Kong
London
Milan
Paris
Tokyo

Adrian E. Scheidegger

Morphotectonics

with 52 Figures

 Springer

PROFESSOR ADRIAN E. SCHEIDEGGER
TU Vienna
Department of Geodesy
and Geophysics
Gusshausstr. 27–29/128
1040 Vienna
Austria

Email: ascheide@luna.tuwien.ac.at

Cover photo: View of the Stauffberg in Switzerland (cf. p. 140 in text) from Lenzburg Castle. Photo
A.E.S. 17 Sep. 2000

ISBN 3-540-20017-7 Springer-Verlag Berlin Heidelberg New York

Library of Congress Cataloging-in-Publication Data
Scheidegger, Adrian E., 1925-
    Morphotectonics/Adrian E. Scheidegger.
        p. cm.
    Includes bibliographical references and index.
    ISBN 3-540-20017-7 (acid-free paper)
    1. Morphotectonics. I. Title.

Springer-Verlag is a part of Springer Science+Business Media
springeronline.com
© Springer-Verlag Berlin Heidelberg 2004
Printed in Germany

The use of designations, trademarks, etc. in this publication does not imply, even in the absence of
a specific statement, that such names are exempt from the relevant protective laws and regulations
and therefore free for general use.

Typesetting: LE-TeX Jelonek, Schmidt & Vöckler GbR, Leipzig
Cover design: E. Kirchner, Heidelberg
Printed on acid-free paper   30/3141/as   5 4 3 2 1 0

Preface

The subject of the present book deals with the relationship between the morphometry and the (neo)tectonics of a landscape. The morphometry refers to the definition of various geomorphological elements by means of numbers; the neotectonics is believed to be mainly indicated by bedrock joints in the area in question. Reasons for this belief will be given in the introductory chapter.

The main purpose of this book is simply to make a comparison between joint sets and landscape elements, mainly with regard to their orientation structures. It has been observed that there is a general correspondence worldwide between the two types of objects, regardless of the age or the origin of the joints. If the joints are indeed somehow of neotectonic origin, there is a significant motive for inferring that geomorphological elements are also to a large extent tectonically co-designed. An additional result is that there is a propensity of geomorphic features (including joints) to be aligned in N-S and E-W directions, which is an indication that the rotation of the Earth must play a fundamental role in their genesis. These conclusions are based on factual data; they are independent of the specific origin of joints, and are independent of any geotectonic hypotheses. However, it is not the aim of the present book to discuss how these results may affect the acceptability of such hypotheses.

The book represents a somewhat personal view of the subject. The vast majority of the joint orientation measurements have been made by the author himself; he has personally visited most of the outcrops on all six continents including the Antarctic and on islands in all major oceans used for this book. Many of the visits were made possible by scientific exchange agreements between the Austrian Government and various countries; for more remote areas, the writer took advantage of field trips organized in the course of meetings by the convening societies, such as of the Geological Society of America to the Antarctic and of the Nepal Geological Society to the Himalayas. The writer wishes to acknowledge his gratitude to these organizations.

Adrian E. Scheidegger

Contents

CHAPTER 1

Fundamentals

1.1
Introduction

"Morphotectonics" deals with the relation of landscape morphology to tectonics. Traditionally, the genesis of many geomorphic landscape features has been ascribed solely to exogenic, nontectonic causes. Thus, drainage systems, the shape of valleys (V-, U-form), incised meanders, glacial forms, volcanic landscapes, mass movements and other features have been attributed to the action of exogenic agents alone. It is the aim of the present book to show that in practically all such cases, a co-design is present due to geotectonic, endogenic causes. Thus, some common notable contentions (e.g. that water causes V-shaped, ice U-shaped valleys) are shown to be false. The proper mechanisms for the genesis of such features are elucidated and many examples are given.

In order to achieve this goal, one has first to discuss the various theories of landscape development as well as the various methods for identifying individual elements therein and describing their features statistically. Next the principal tectonic agents and their manifestations have to be discussed and related to particular morphological features. Finally, the significance of correlations has to be established.

1.2
Landscape Development

1.2.1
General Considerations

Landscapes display aspects of bewildering complexity. Nevertheless, one can recognize a series of principles (Scheidegger, 1991) which are evidently operative in their genesis and evolution. These principles are fundamentally of two types: *phenomenological* and *theoretical*. The phenomenological principles have been abstracted (so to speak "inductively") from observations; the theoretical principles are basically mathematical models based on complex systems theory. Amongst the latter, one has to distinguish further between *dissipative* and *self-ordering* systems.

1.2.2
Phenomenological Principles

1.2.2.1
General review

A summary of the phenomenological principles of landscape evolution has been produced by Scheidegger (1987). Accordingly, the basis of such principles had implicitly already been recognized by Davis (1924) who thought that landscapes develop cyclically from a youthful stage, initiated by endogenically (i.e. originating from "inside", Greek "endo", the solid Earth) induced tectonic processes, through maturity to old age, owing to the continuous action of exogenically (i.e. originating from "outside", Greek "exo", the solid Earth) induced erosion. Strahler (1957) quantified this idea by introducing the concept of a *hypsometric curve*: the latter is obtained by calculating the fraction of the area of the landscape under consideration that lies below a certain height level. In geomorphology, it is common practice to consider *relative* hypsometric curves: The heights and areas are divided by the total height (difference between the highest and lowest point in the area) and by the total area, respectively. Then, Strahler (1957) assumed that the hypsometric curves are convex for "youthful" landscapes, ±straight for "mature" landscapes and concave for "old" ones (Fig. 1.1). This "cycle" theory of Davis is now no longer tenable since it has been recognized that endogenic and exogenic processes can occur concurrently: although no erosive degradation can take place without a tectonic buildup, endogenic buildup and exogenic degradation generally act concurrently as antagonistic processes (Scheidegger 1979e).

Thus, what remains today of Davis' cycle theory is an *Antagonism Principle* which is the most fundamental principle of landscape evolution. It refers to the fact that there are essentially *two* types of processes active in the formation of

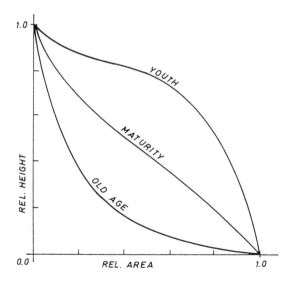

Fig. 1.1. Hypsometric curves for young, mature and old landscapes (modified after Strahler, 1957)

a landscape at the same time: the endogenic processes originating from inside and the exogenic ones originating from outside of the solid Earth. Endogenic and exogenic processes occur *concurrently*; they act "antagonistically" at the same time. Closely related to the Antagonism Principle are several important corollaries, viz. the *Instability Principle* and the *Catena Principle*. In addition, the endogenic processes are responsible for some stress-related landscape principles; viz. the *Selection Principle* and the *Tectonic Predesign* (or *Co-design*) Principle. We shall discuss these principles in detail below.

1.2.2.2
Antagonism Principle

The implications of the Principle of Antagonism itself have been analyzed in depth by Scheidegger (1979e). It can be summarized as follows: The *nature* of a landscape, its *development stage* in the sense of Davis (1924), i.e. whether it is in the Davisian stage of youth, maturity or old age, is not determined by the developmental sequence at all, but by the intensity of the antagonistic processes. Quite generally, when in dynamic equilibrium, the uplift (v_u) and denudation (v_d) rates (they are velocities) balance each other more or less in a landscape. If this rate is high, i.e. > 0.5 mm/a, the landscape presents aspects of "youth" (*youth*-curve in Fig. 1.1), if this rate is medium (± 0.5 mm/a), the landscape is in the stage of "maturity" (*maturity*-curve in Fig. 1.1), if the rate is low (< 0.5 mm/a), the landscape is in the stage of "old age" (*old-age*-curve in Fig. 1.1). If the uplift velocity v_u is not equal to the denudation velocity v_d, non-stationary conditions prevail. One can determine a "stationarity index" S by taking the ratio

$$S = v_u/v_d \qquad (1.1)$$

which is 1 under stationary conditions. If $S < 1$, the landscape decays, if $S > 1$, it is being built up. In stationary landscapes, the relief follows Strahler curves. In non-stationary cases, the stage is shifted to a "greater" age if the uplift is greater than the denudation. Adams (1980) has shown that peaks are flat-topped if the uplift rate exceeds the denudation rate, and spiky if the uplift rate is equal to or smaller than the denudation rate.

The two processes involved in geomorphic antagonism have fundamentally different stochastic natures: exogenic processes are essentially random, endogenic ones essentially non-random (Scheidegger 1979e). The recognition of this fact is based on induction from a vast amount of observational data.

The randomness of exogenetic processes results from the evidently "chaotic" nature of weather phenomena: The meteoric activity affects a particular landscape more or less randomly, although the general climatic trends may, of course, be more systematic.

The situation in the case of endogenetic processes is exactly the opposite: These processes are the consequence of tectonic motions which are caused by a global neotectonic stress field. The latter is systematic over large (plate-wide) distances and therefore causes systematic features on a correspondingly large scale. In this fashion, many geomorphic features, such as valley trends, the genesis of gorges, the creation of piedmont lakes etc. can be explained as endogenetically predesigned (cf. the section on *Tectonic Design Principle* below).

As noted in the introduction, the Antagonism Principle has a number of important corrolaries; these will now be discussed.

1.2.2.3
Instability principle

The idea that antagonistic processes roughly balance each other in a stationary dynamic geomorphic state has to be modified by the observation that this "balance" is quite often unstable (Scheidegger 1983a). Apart from the fact that individual surface features tend to be impermanent, even though their overall appearance may seem to be constant, the dynamic equilibrium may actually be inherently unstable: Any deviation from uniformity tends to grow, and there is, so to speak, a positive feedback between the size of the deviation that has already been reached and the rate by which this size increases. The formalism to analyze instabilities in geomorphology was probably initiated by Slingerland (1981) who investigated (random) deviation in a dynamic landscape system. Although the dynamic equations of the system are unknown, the (presumably) nonlinear differential equations can be linearized for the purpose of stability analysis according to Taylor (1950), at least in the vicinity of the deviation for a short time range. Then, for any parameter $x(t)$ describing a landscape feature, the solution of the differential equation is an exponential function of time and thus the growth equation at the beginning has the form

$$x(t) = C \exp(\lambda t) \,, \tag{1.2}$$

in which C is a constant. If λ is positive, the condition is unstable, the deviation $x(t)$ grows exponentially and the value of λ is a measure of the degree of the instability. The exponential function would increase infinitely so that it would, by implication, lead to a catastrophic situation. However, the linearization holds true only for short time ranges so that an unstable state need not necessarily lead to a catastrophe: the growth process may eventually come to a stop when a saturation stage is reached.

The Instability Model is evidently correct for many purely exogenic geomorphic features: The longitudinal profile of stepped valleys, the terraced transverse profiles of valleys, sequences of pools and riffles in river beds, the meandering of rivers, the formation of glacial cirques and certain dissolution phenomena (such as "onion-skin" weathering). Some of the above features can also be described as a consequence of a "Catena Principle", which is very closely related to the Instability Principle. It will be treated next.

1.2.2.4
Catena Principle

Statement   The Catena Principle is very closely related to the Instability Principle. We have already mentioned above that the latter leads to such features as terraced valleys and sequences of riffles and pools in river beds. In effect, such sequences are "chains" (Latin "catenae") of geomorphological elements that are repeated again and again; this is the consequence of instabilities that arise at certain points but reach a "saturation stage" so that the corresponding features are repeated in a sequence. This can be expressed as a separate principle, viz. the *Catena Principle* (Scheidegger, 1986). The concept of a catena was originally devised by Milne (1935, 1947) in soil research who noted that certain definite sequences (catenas) of soil types recur on a slope; it was then adopted generally for self-repeating morphological features. In particular, the Catena Principle has been applied to *hydrological* elements on "slopes" (this includes the course of a river or gully). The fundamental elements of a catena are best observed on a single, solitary, free slope. Fundamentally, a catena (Fig. 1.2) encompasses an *eluvial* region (I) at the top (of flat topography with a lip), a *colluvial* region (II) in the middle (of steep topography with large mass flow rates) and an *alluvial* region (III) at the bottom (again of flat topography). The catena principle then states that the whole landscape is made up of catenas each of which consists of a sequence of flat-steep-flat sections.

Theoretical considerations   One can ask oneself how the basic catena form can be explained in terms of mechanics. In effect, the genesis of a "catena" is a consequence of fundamental system dynamics; the geomorphological activity is proportional to (increases with) the topographic relief. This condition can be formulated in mathematical terms: Let y be the topographic height, x the (lin-

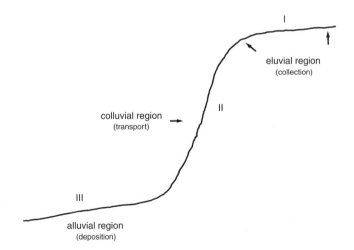

Fig. 1.2. Catena with (*I*) eluvial, (*II*) colluvial and (*III*) alluvial regions (after Scheidegger, 1986)

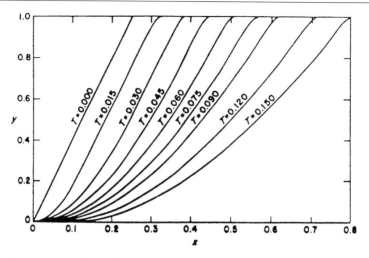

Fig. 1.3. Numerical solutions of Eq. 1.3 for a sequence of times T in arbitrary units (after Scheidegger, 1961)

ear) spatial coordinate and let the two original levels (eluvium and alluvium) of a catena ($y = 0$ and $y = 1$) be connected by an inclined straight line. Then, if the rate of change (erosion) is proportional to the declivity of the slope, the differential equation governing the process is (Scheidegger, 1961)

$$\partial y/\partial t = -(\partial y/\partial x)\sqrt{[1 + (\partial y/\partial x)^2]} \, . \tag{1.3}$$

Numerical solutions of this differential equation are shown in Fig. 1.3 for a sequence of times T in arbitrary units. The patterns in this figure describe the basic morphology of a catena. We shall now deal with some specific applications.

Slope catenas   We have already mentioned above the fundamental aspect of a slope catena: It consists of a flat region, followed by a steep one, ending again in a flat one. This statement refers to a single, solitary catena. These are found only occasionally in nature: slopes are seldom solitary. Generally, they are coupled with other features. A common case is that of a slope being located at the side of a valley and then "coupled" with a river course (Fig. 1.4). In such cases, the eluvial zone (I) is still present, but the colluvial zone (II) may already be interfered with (II$_m$) and the alluvial zone (III) may be missing altogether due to the erosive action of the river.

Fluvial catenas   A fluvial course (thalweg) may be considered as consisting of catena elements. The basic tripartite structure of a fluvial catena was first recognised in mountain torrents by Rehbock (1929). It is composed of an eluvial slope (this is the catchment area with erosion cirques and collection flats for the slope water), the colluvial zone (this is a narrow valley sector

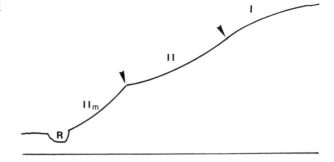

Fig. 1.4. Catena coupled with a river (after Scheidegger, 1986)

in a mountain stream, generally of the appearance of a small gorge) and an alluvial zone (generally an alluvial cone). In nature, this scheme may repeat itself again and again and lead to the well-known alternation of broad-flat and narrow-steep sections of Alpine valleys.

Mass flow catenas   Such catenas are encountered in mud- and debris-flows. The latter refer to the motion of water-saturated material which comes into a state of flowage. As a general observation, it may be noted that a mass-flow catena can never start at a watershed, because the material has to be saturated by water entering from above so that it can move. Thus, there must be a catchment area above the catena. The basic scheme is shown in Fig. 1.5. A typical case of such a catena has been described by Gerber and Scheidegger (1984) from a famous clay slide near Schinznach in Switzerland.

Scree catenas   A special type of mass-flow catena is found where a rock wall stands above a slope. If the transition from the wall to the slope is connected with a transition of the country rock from a firmer to a softer consistency, then the wall-material will scale off and a scree slope forms below the wall. At this point, a catena begins with the debris accumulation; the "eluvium" refers now to the collection of scree. The colluvial zone is the transport zone of the scree slope and the alluvial zone is the zone of deposition of the scree (Gerber and Scheidegger, 1974).

1.2.2.5
Selection Principle

Statement   Totally different types of phenomenological landscape development result from mechanical stresses. Some of the most striking features in landscapes are single towers, tors and pillars: Gerber (1969) has noted that these features have forms that are inherently statically stable under the action of their own weight. He enounced this as the *Selection Principle* which refers to the observation that the exogenic processes of the Antagonism Principle operate under certain circumstances in a directed fashion without an actual external (exogenic) influence being present. It states: *during the weathering processes in a landscape, those morphological elements remain that ensure the*

Fig. 1.5. Basic scheme of a mass-flow catena (after Scheidegger, 1986)

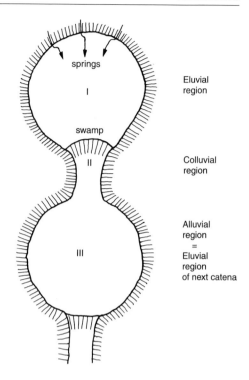

springs

I

Eluvial region

swamp

II

Colluvial region

III

Alluvial region
=
Eluvial region of next catena

static stability of the resulting features. The formation of statically favourable forms is, thus, the result of a natural selection process in the sense that of the multitude of forms created by erosion, those remain for the largest period of time which are the most stable with regard to the self-gravitational (not tectonic) geostatic stresses.

Phenomenological confirmations Phenomenological confirmations of the operation of the Selection Principle can be seen in the existence of various types of *pillars*, *bastions*, and *towers* that are found in many parts of montainous regions. The *mesas* and *buttes* also attest to the operation of the Selection Principle, in as much as their forms are exactly the geostatically stable ones. Other features belonging to those possibly influenced by the Selection Principle are *garlands* of heads on escarpments: a particularly striking example of such features has been described in the Jura mountains of Switzerland by Scheidegger (1985b). The principle may also be operative in the decay of *rock walls*. Here the decay basically proceeds from the foot of such a wall upwards (Gerber, 1963; Gerber and Scheidegger, 1965).

Mathematical Theories Analytical theories of the genesis of the mentioned features based upon the Selection Principle have been given on several occasions. Thus, it is easy to show that a pillar with a broad base and a narrow top is a stable feature. In fact, if the bounding contour line is an exponential curve, the pressure due to the weight of the overlying mass is the same at every level

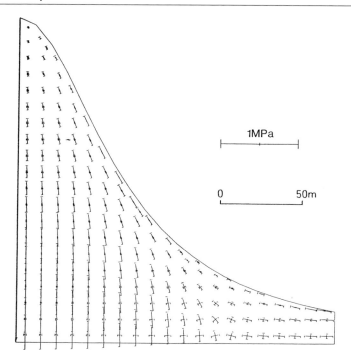

Fig. 1.6. Principal stresses (pressures) in a symmetrical crest (modified after Scheidegger and Kohlbeck, 1985)

in the pillar which makes the latter a stable feature (cf. the Eiffel Tower in Paris). If the cross-sectional area at a height h above ground is denoted by πr^2, r being the radius at height h, then the pressure p at that height is (Gerber and Scheidegger, 1975)

$$p = \int_{h}^{\infty} \pi r^2 \, dh \, \rho g / (\pi r^2) \, , \tag{1.4}$$

where ρ is the density of the material and g the gravity acceleration. Assuming for r an exponential function of h

$$r = A \exp(-ah) \tag{1.5}$$

(with A, a constants) indeed yields a constant pressure

$$p = \rho g / (2a\pi) \, . \tag{1.6}$$

This proves the contention presented above.

A corresponding calculation of the self-gravitational stresses in a *ridge* has been made by McTigue and Mei (1981), based on the method of complex potentials and using rather severe simplifications. A proper calculation relies on

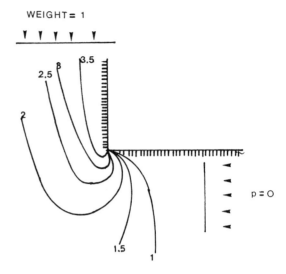

Fig. 1.7. Isolines of p_{max} near the foot of a wall (after Sturgul and Scheidegger, 1967)

finite element procedures carried out on a computer. No general solutions can be obtained in this fashion; however, Scheidegger and Kohlbeck (1985) have calculated some examples. In Fig. 1.6 we show the principal stresses (pressures) in a symmetrical crest (only right side shown) as calculated by these authors. The crest has no tensile stresses in its centre and shows only low stress concentrations at its base. Therefore, it appears to be a stable configuration and thus suitable for selection during the weathering processes.

Rock walls, as was mentioned above, always decay from the bottom upward. This fact is directly due to the action of the self-gravitational stress field. At the foot of the wall, stress concentrations occur which initiate the decay; furthermore, a straight wall is a stable configuration, since protrusions would be sheared off. Analytical calculations of the stresses in such a wall have been made by Sturgul and Scheidegger (1967) by finite element techniques. Figure 1.7 shows the isolines of p_{max} near the foot of a wall. The stress concentrations become clearly visible which explains the selective weathering occurring in this region.

1.2.2.6
Tectonic Design Principle

Another instance of a mechanical landscape principle is the *Tectonic Design Principle*. It states that not only the self-gravitational stresses, but also the deep-seated tectonic processes and the attendant stresses can influence geomorphic features in a fundamental way. Although geomorphologists have generally sought the origin of many of today's landscape features in exogenic processes operating within the framework of the Antagonism Principle (rivers are supposed to have cut gorges through rising mountains retaining their antecedent courses, glaciers are supposed to have widened valleys from a V to a U shape, and landslides are supposed to be caused solely by exogenic agents), this contention is obviously false (see e. g. Scheidegger and Ai, 1986; Hantke, 1991 & 1993). The main purpose of this book is to present the corresponding arguments.

1.2.3
System Theory

1.2.3.1
Dissipative systems

General remarks   Thus, the evidence has been increasing that exogenic effects can be adequately described by the (quasi-) random action of exogenic agents. This has been shown to hold true for the meander formation, the surface erosion, the drainage basin development in unstructured plains and the large-scale decay of slope banks and mountain ranges (Scheidegger, 1979e). This is also the reason that it has been possible to describe landscape evolution by the application of *Stochastic General Systems Theory* which was developed originally by Bertalanffy (1932) for applications in connection with biological questions. In order to speak of a system, one needs (i) a set of elements identified with some variable attributes of objects, (ii) a set of relationships between attributes and objects, and (iii) a set of relationships between attributes of objects and the environment. Thus, "a system is a set of interrelated elements which function together as an entity embedded in an environment" (Harvey, 1969). The last condition assumes that the system is open to some external environment, which, however, is not necessarily the case.

The *Formal System Theory* is, in fact, a mathematical discipline. It is generally involved with statistical ("stochastic") methods. In geomorphology (Carso and Kirkby, 1972), one deals basically with mechanical systems: The objects are landforms (slopes, rivers, coasts), the attributes are quantifiable properties of these landforms (e. g. drainage density, slope declivity, meander curvature) and the "relationships" consist of mechanical exchanges of energy and mass between the landforms so that causal links are formed between the attributes. The distinction between the system itself and the "environment" is arbitrary and made by the observer. A natural delimitation occurs only in the case of systems that are completely closed; geomorphic systems, however, are not normally closed. The state of the system is defined by giving all the attribute values of all the elements; the set of all attribute values can be represented by a point in a multi-dimensional phase space. During the evolution of the system, this point will describe a trajectory in phase space. The number of elements is large and therefore the attribute values of each element cannot be ascertained in detail in practice. Therefore, there is a quasi-probability distribution of phase points which represent the position-likelihood of the system in phase space within the limits of one's knowledge. This probability distribution can be taken as the basis for statistical predictions of the behavior of the system. Within the limits of one's knowledge, a whole "ensemble" of states is possible. Because of some fundamental natural laws, e. g. conservation of mass (or energy), not all thinkable states in phase space are possible; these may be restricted to certain regions. An observable quantity is generally a crude characteristic built on a conglomerate function of attributes. The expected value for this observable quantity is the average of the conglomerate over all states of the system that are possible. On occasion, the conglomerate function has been calculated for those attributes that correspond to the most probable state of the system, but this is not in conformity with the principles of statistical physics as they were developed by Boltzmann and Gibbs in connection with gas dynamics (see e. g. Sommerfeld, 1964). Thus, the "most probable" and the "expected" characteristics are not the same; it is logically evident that ensemble averages have to be taken for the attributes, not those attributes for the most probable state of the system.

In *equilibrium conditions* (*Ergodic Principle*) the basic probabilities stay stationary. In phase space, this is expressed by the confinement of the phase points to certain fixed regions. If there is a constant of the motion H, this region is represented by the subspace that has the equation

$$H = \text{const}. \tag{1.7}$$

If the interactions are numerous and complex, the phase point will move in time; because we suppose equilibrium conditions, all points of the subspace will, in time, be equally closely approached by the phase-point: This is the ergodic theorem embodying the idea that the time averages can be replaced by ensemble averages (cf. e. g. Sommerfeld, 1964). If the system consists of a large number of subsystems which fluctuate and contribute part values H_i to the constant of the motion H, then the distribution P_i of these values in canonical (cf. e. g. Sommerfeld, 1964) form is

$$P_i(H_i) = (1/Z) \exp(-H_i/kT), \tag{1.8}$$

where Z is the partition function required for normalization and kT a parameter. This is a consequence of the limit theorems of probability theory under the assumption that the interactions between subsystems are numerous and uncorrelated. On this basis, it is possible to set up a complete analogy between thermodynamics and landscapes; one can define analogs of all the thermodynamic functions (notably the entropy) and corresponding landscape variables (Scheidegger, 1967a). It is clear that the constant of the motion is mass (this is conserved in the landscape process), the temperature equivalent is the topographic height h, and for the entropy analog S one has

$$dS = dM/h. \tag{1.9}$$

The analogy between temperature T and topographic height h and the corresponding definition of "geomorphic" entropy S had been postulated originally by Leopold and Langbein (1962) on entirely heuristic grounds, from the analogy with a heat engine. These authors observed that the thermodynamic principle applied to the landscape analogs of the thermodynamic variables led to valid statements regarding landscape development. The statistical justification of these analogies lends a much better foundation of the latter.

Incidentally, a "landscape entropy" has also been defined independently of the above considerations by cartographers (Zdenkovic, 1976). It can be shown (Lechthaler-Zdenkovic and Scheidegger, 1989) that it is the same as that defined by Leopold and Langbein (1962). If the analogy between thermodynamics and landscape development theory is applied to steady states, for which the entropy production rate σ must be a minimum of

$$\sigma = - \int (h'/h^2) J \, dx = \min .$$ (1.10)

in which J is the mass flux per unit time, then it is possible to calculate equilibrium river profiles, etc. The equilibrium theory immediately leads to the process–response concept: as soon as an equilibrium is disturbed, the system responds by the adjustment of the remaining variables to a new equilibrium configuration.

The analogy between thermodynamics and geomorphic systems can be extended to nonequilibrium cases and then leads to the *Principle of Dissipative Degradation*. If the (large) system possesses a large number of subsystems in which a non-negative quantity is transferred by a statistically fluctuating transfer process whose exact nature is unspecified, then the central limit theorem of probability theory leads to a diffusivity equation for the non-negative quantity, provided the fluctuations are linearly additive (Tomkoria and Scheidegger, 1967): in thermodynamics the temperature is subject to a diffusivity equation

$$\partial T/\partial t = D[(\partial^2 T/\partial x^2) + (\partial^2 T/\partial y^2)] .$$ (1.11)

In view of the analogy between geomorphic height and temperature a corresponding diffusivity equation can be postulated for the height h in a landscape

$$\partial h/\partial t = D[(\partial^2 h/\partial x^2) + (\partial^2 h/\partial y^2)] .$$ (1.12)

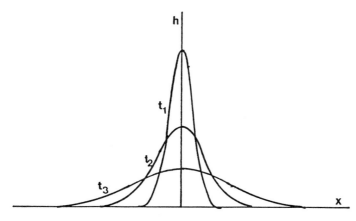

Fig. 1.8. Qualitative sketch of the solution of the diffusivity equation showing the dissipative decay of a peak with time t (initial form: delta-function)

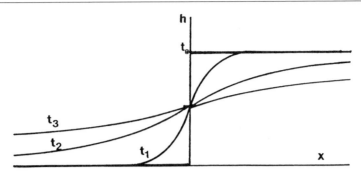

Fig. 1.9. Qualitative sketch of the solution of the diffusivity equation showing the dissipative decay of a slope bank with time t (initial form: rectangular slope bank)

For the diffusivity equation, one has the following well-known solutions (in one dimension)

$$h = (4\pi Dt)^{-1/2} \exp[-x^2/(4Dt)] \,, \tag{1.13}$$

$$h = \tfrac{1}{2} + \tfrac{1}{2} \operatorname{erf}[x/(4Dt)^{1/2}] \,. \tag{1.14}$$

These are equations that describe the decay of (the profile of) a peak (Eq. 1.13; Fig. 1.8) and of a slope bank (Eq. 1.14; Fig. 1.9). Thus, the theory immediately leads to an understanding of the phenomenon of dissipative degradation.

The quasi-stochastic system approach implies the existence of a *Principle of Catastrophic Thresholds* (Thom, 1972). When one or more parameters possess one or several critical "bifurcation" values at which small changes of the control parameter lead to further hierarchically arranged bifurcations, these may be run through until a quasi-stationary state of complete chaos is reached (Schuster, 1984; Brun, 1986; Harrison and Biswas, 1986).

The theoretical models based on the dissipative system theory correctly describe the decay of landscape features and the possible occurrence of "catastrophes". However, most landscapes are not decaying, but represent a dynamic quasi-stationary state. For this, one must use models based on open complex systems as outlined in the next sections.

1.2.3.2
Self-Ordering Systems

The system theory, as described and applied above, is essentially based on the operation of the Ergodic Principle and the decay of closed dissipative systems. However, the Ergodic Principle evidently does not generally apply, inasmuch as the phase trajectories often do not range through the whole phase space, but are concentrated in the neighborhood of *attractors*. In effect, complex systems often develop some internal order spontaneously. The pertinent questions have been investigated in a new branch of science, called *complexity theory*.

In nature, there are innumerable examples of relatively stable self-organized ordered states at the edge of chaos (Bak et al., 1988). This applies particularly to landscapes: An ordered state at the edge of criticality establishes itself in a complex landscape system solely on account of the (normally highly non-linear) interactions between the individual elements of the system and not because of the presence of an external ordering principle.

The background of the above is the description of the natural systems in terms of the complexity theory. It is noted that the ultimate evolution of the system is determined by the attractors to which the trajectories converge. To define attractors in mathematical terms (Monin, 1991), one has to define *non-wandering phase points* whose neighborhoods intersect some phase trajectory at least twice; *invariant sets* (of phase points) which are filled with whole trajectories, and *minimal sets* which are nonempty closed invariant sets having no subsets with the same properties. Then, *attractors* are minimal sets Λ of nonwandering phase points having neighborhoods in which all the trajectories approach Λ asymptotically; attractors differing from stationary points and limit cycles are called *strange*.

The recognition of the frequent presence of self-organized criticality in many systems resulted from sheer observation. Thus, with reference to geomorphology, one found that the characteristic observables in such quasi-stationary, ordered states as represented by the distribution of heights in a landscape or by the distribution of mass in a sand pile (Bak et al., 1988) are generally spatially and temporally scale-invariant; they have been found to be fractal. In a fractal set of dimension D, a power law exists for subsets: The number N of subsets of (linear) "size" L is proportional to L^{-D}. Stationary, ordered states seem to occur only at the edge of criticality. This might be a general feature of nature (Nicolis and Prigogine, 1977; Lundqvist et al., 1988; Haken and Wunderlin, 1991; Cramer, 1993; Kauffman, 1993). Thus Bak et al. (1988) noted that *certain extended dissipative dynamical systems naturally evolve into a self-organized critical state; the temporal fingerprint of such states is the presence of flicker [or red] (= 1/f) noise* (cf. Dutta and Horn, 1981), *its spacial signature is the emergence of a scale-invariant (fractal) structure.*

The spontaneous establishment of ordered conditions at the edge of chaos has been simulated by computers for innumerable examples of nonlinear systems. However, computer simulations can only emulate the development process *a posteriori:* No *reason* can thereby be brought forward why the development should occur as it does. Therefore one would like to have more *fundamental* insights into the circumstances that govern the establishment of self-organized order at the edge of criticality. One can, first of all, look for conditions that are necessary for the spontaneous establishment of a quasi-stationary, ordered state.

Landscape systems are evolutionary in certain respects: Elements are "born" and "die". Evidently, for a stationary state to develop, the death rate must be equal to the growth rate in a system, otherwise one has complete obliteration of the system or an explosion. One can see that this is a critical condition.

Furthermore, all "evolutionary" systems are open and dissipative. Therefore, the laws of equlibrium thermodynamics do not hold: non-equilibrium thermodynamics have to be applied. The usual form of the second law of thermodynamics does not apply: in open systems the entropy may well decrease during the approach to a steady state, i.e. its value at the nonequilibrium steady-state may well be smaller than at equilibrium (Prigogine, 1947). In linear theory, a theorem of minimum entropy production holds for stationary states. Any perturbation in such a stationary state always regresses; the steady states near equilibrium are essentially uniform in space if permitted by the external constraints. The stability of these states implies that *in a system obeying linear laws, the spontaneous emergence of order in the form of spatial or temporal patterns differing qualitatively from equilibrium-like behavior is ruled out.* Moreover, any other type of order imposed on the system through the initial conditions is destroyed in the course of the evolution to the steady state (Nicolis and Prigogine, 1977, p. 46). Thus, a necessary condition for the spontaneous development of order is that the relationship between the elements of the system under consideration is nonlinear.

Next, one can ask oneself why a stationary state at the edge of chaos has to be fractal. We have seen above that "order", by its definition, requires the corresponding attractor to have a dimension which is much smaller than the dimension of the entire phase space; this prevents the system from wandering all over the entire phase space (= complete chaos) and confines it to a small region thereof (= "relative" order or "low-dimensional chaos"; cf. Kauffman, 1993, p. 178–179). However, there is no condition that this dimension be an integer: If it is not an integer, it is fractal (in the sense of Hausdorff, 1919) – there are no other possibilities.

The order at the edge of chaos has to be *stationary* (at least for a while) with regard to small changes in the initial conditions. The latter are generally expressed by changes (fluctuations) of the *parameters* which can cause instabilities within the attractor to occur or can cause the system to move to an entirely different attractor. Instabilities within the (strange) attractor can be caused by accidental fluctuations or by random external forcing. They have often been referred to as the "butterfly effect": A butterfly batting its wings in Peking stirs the air and thereby causes instabilities in the system which magnify so intensely that a storm develops in New York (Lorenz, 1963; Çambel, 1993, p. 194). In mathematical terms, this is the result of a Lyapunov instability (Monin, 1991): The (small) initial distance between trajectories on the attractor in question increases exponentially with time (this is the case with the Lorenz [1963] attractor modeling weather systems). The corresponding coefficient is called "Lyapunov coefficient" (dimension T^{-1}): If the latter is positive, the system is unstable. A more drastic instability occurs if the change in parameter values forces the system from being under the influence of one attractor to that of another; this represents a "catastrophe" in the sense of Thom (1972). Thus, it is observed that ordered states are the *only* ones that have duration and that the problem of why order develops out of chaos has been solved (Scheidegger, 1996).

Now it remains to apply this general theory to *landscapes*. We have already mentioned the relative stability of the height-distributions in a landscape:

except for the principle of "self-order at the edge of chaos" there would be no reason why a region like the Alps could present a relatively stable aspect. The morphology is that of an open, self-regenerating system. Noting that the erosion/uplift rate in tectonically active regions is in the order of several mm/a (1−2 mm/a in the Alps [Gubler, 1976], up to 10 mm/a in the Himalaya [Kisaki, 1994]), i. e. several km/Ma, it is clear that the whole structure would change its aspect completely in a few Ma if it were not for the self-ordering principle that has preserved e. g. the "mountain range aspect" of the Alps at least since the Miocene.

1.2.4
Integrated View of Landscape Development

When we review the landscape principles outlined above, we see that there are in fact two different classes: the first class of principles refers to abstractions from phenomenology, the second class to predictions of mathematical system theory. In the latter, one has to distinguish further between dissapative and fully complex systems. The phenomenological principles are simply inferred inductively from observations, the system principles depend on the correctness of the underlying a priori assumptions. Thus, the phenomenological principles are tentative hypotheses gleaned from observations that can be assumed to be valid until "falsified" in the sense of Popper (1984), whereas the system theory principles are based on probabilistic models, assuming the presence of many elements and uncorrelated relations so that averages can be obtained and the assumption of quasi-stochasticity holds. The development of order in complex systems at the very edge of chaos seems to be a general law of nature: the reason is because ordered states are the *only* ones that have any duration; therefore, these are the patterns that are observed in a landscape.

The combination of the various types of landscape principles leads to an *integrated* view of landscape development. Whilst the original view was that the latter is conditioned by individual specific mechanical processes, the new integrated view assumes that a landscape is an *open, nonlinear complex system* more or less in a dynamic equilibrium. It is governed by a fractal attractor (self-ordered state) which can easily become chaotic ("butterfly effect", Thom's catastrophe). The *input* into the open system originates from neotectonic processes which are systematic in nature; they are primary and most important (without tectonic uplift no degradation!) and lead to tectonic predesign of well-nigh all landscape features. The *output* is represented by random exogenic degradation and erosion. This is a secondary feature, following the primary uplift. Generally the primary tectonic predesign is not obliterated: Thus, drainage systems (river valleys and gorges) are *not* due to the erosion by antecedent rivers. Previously existing rivers *follow* tectonic lines (see section on tectonic design principle); the primary design is not randomized or superseded by exogenic erosion. Consequently, no landscape is static; there is a constant evolution, much like in living beings. A landscape, therefore, can principally not be stabilized, much to the chagrin of engineers who whish to create stable conditions on slopes, river banks, meander trains, etc.

1.3
Orientation Studies

1.3.1
Significance

In morphotectonic studies, the *orientation* of various elements (such as joints, lineaments, rivers, crest-lines etc.) are of essential importance. If the orientations of such features correlate, it can be hypothesized that they have a common origin. The orientations of geomorphic elements are commonly represented by *axes*, which are defined by their azimuth (angle from the northern direction measured eastward), and in addition, by their dip (or "plunge") angle (angle in degrees from a horizontal plane). The orientation of a plane is best given by the azimuth and plunge angle of its polar axis (axis standing normal on the plane); an alternative method is that of giving azimuth and plunge of its dip direction (line of steepest descent) or its strike direction (azimuth of the intersection line of the plane with a horizontal plane) and the dip angle (plunge of line of steepest descent). In the last instance it must be stated whether the plane dips to the right or to the left of the strike–azimuth direction. In many instances, either the plunges of axes are very shallow ($\pm 0°$) or the dips of surfaces very steep ($\pm 90°$), so that the pole or the strike directions alone are sufficiently significant.

In any one area, the orientation of certain types of geomorphic elements (such as joints, river segments, ridges) are commonly grouped into clusters or sets. Within each set, the orientations are very similar, but not identical. Thus, one is faced with the problem of reliably determining the "mean" orientation embodied by one or several clusters of axes, which can only be solved by statistical methods.

1.3.2
Nonparametric Statistics

The usual procedure for the statistical analysis of axes is non-parametric: One plots the axis density or axis rose (in the latter case ignoring the dips/plunges) diagrams and visually handpicks the peaks. For this purpose, an equal-area ("Lambert projection") of the unit sphere has to be used; it is then usual to plot the intersection points of the axes with this sphere and to state the frequency of such points in "percent (%)": Assuming that a total of 100 measurement points exist, this is the number of points present within a certain "orientation window" whose circular area is 1% of the area of the lower half of the unit sphere (method of Müller-Salzburg, 1963). It is then usual to draw isolines of equal density on axis density diagrams; the maxima above an arbitrarily fixed cutoff are then taken as "prevailing" or "mean" orientations. No assumptions are made with regard to the distribution that is expected with regard to the points; the cutoff is simply raised until a few maxima become evident. Computer programs have been devised to draw the isolines automatically; the maxima are handpicked from the plots. If the dips (plunges) are of no interest, corresponding rose

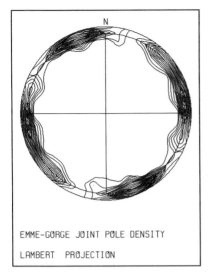 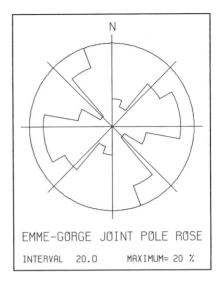

Fig. 1.10. Pole density diagram (*left*) and pole rose (*right*) of a set of 63 joints measured in the Emme Gorge in Switzerland (note that, in order to obtain strikes, the diagrams would have to be turned by 90°)

diagrams are drawn, for which the counting intervals are adjusted until a few peaks become evident. Again, the procedure has been computerized. Figure 1.10 shows the pole density and pole-trend of a typical joint-set.

The question then arises as to how many data are necessary for getting "reliable" results. Some structural geologists claim that thousands of measurements of the pertinent data have to be taken; others think that a few tens or hundreds are sufficient. This question cannot be solved by non-parametric statistics.

1.3.3
Parametric Statistics

Parametric statistics are based on the assumption that a particular set of data has a distribution which is described by a pre-assumed function with a number of unknown parameters. This assumption is usually not really satisfied, but in practice it can be fulfilled very closely in many cases.

With regard to orientation studies of the axes, one generally tries to make a best fit of the data referring to one cluster to a theoretical so-called "Dimroth–Watson" probability distribution (Dimroth 1963, Scheidegger 1965, Watson 1966) which, on a sphere, corresponds to the Gaussian distribution on a plane. It is described by the equation

$$f(\theta) = A \exp(K \cos^2 \theta) , \qquad (1.15)$$

where θ is the deviation angle from a central axis, given by a unit vector refer-ring to the distribution (see Watson 1966), and where A and K are parameters (Kohlbeck and Scheidegger 1977, 1985). This method, when applied to a prac-tical problem, consists in the determination of the "best-fitting" parameters of the theoretical distribution by minimizing the sum of the square deviations of the observed from the predicted values: Let $x(x, y, z)$ be the direction vector of the sought "best" axis, and $n_i(n_{ix}, n_{iy}, n_{iz})$ the direction vector of the i-th measured axis, then the function F

$$F(x, y, z) = \sum (x.n_i)^2 \tag{1.16}$$

(the dot denotes the scalar product) must be a minimum subject to the condi-tion that $x^2 + y^2 + z^2 = 1$ (the sought axis must be a unit axis). This is a well known problem concerning quadratic forms and can be solved by solving a third order equation (Scheidegger 1965). The final Dimroth–Watson distri-bution is characterized by 4 parameters: the direction of its central axis (two parameters), the value of K and that of A, but since the integral of the distri-bution over the lower half of the unit sphere must be 1, only three parameters are independent.

Extending this to N clusters, one notes that, although each distribution is characterized by four parameters, the integral of all of them over the unit sphere must again be equal to 1, so that N distributions do not require the determination of $4N$, but only of $4N - 1$ parameters. (This refers to spacial axes such as representing joints; in case of linear data representing rivers or ridges, each individual cluster is defined by 3 parameters; thus N distributions require $3N-1$ parameters.) These parameters must be defined in such a fashion that the likelihood of finding the actually found data becomes a maximum (or the mean square deviation of the observed from the predicted values a minimum). The solution of this problem can no longer be found by solving a third order equation, but it must be found by directly maximizing a suitable likelihood function first by a Monte Carlo search and then finding the actual maximum by various approximation procedures, which are carried out by specially developed computer programs (e. g. Kohlbeck and Scheidegger 1977) which not only give the "best" mean directions, but also parameter error estimates at the level of 0.1 (90%).

One can then address the question regarding the number of measurements that are necessary to determine the "mean" axes of clusters of axes. In the geoscientific literature one finds, as noted, estimates from hundreds to thou-sands of individual orientations that are necessary for the fixation of the mean directions. However, it is a well known fact that around three measurements suffice for the determination of each parameter. The intuitive reasoning for this is as follows: One measurement may be spurious, of two measurements one may be spurious, thus one needs three measurements to fix one parameter by taking their mean. Most testing of standards, e. g. for testing the strength of a block of concrete, (OeNORM, DIN) require three samples to be taken to obtain "reliable" values by averaging. Thus, one requires 9 for a single axis (3 parameters), 21 for 2 axes (7 parameters), and 33 measurements for 3 axes

(11 parameters). This applies to joints; for linear data representing rivers or ridges, 6 measurements for a single axis (2 parameters), 15 for two axes (5 parameters), and 24 measurements are required for 3 axes (8 parameters). Taking a greater number N of measurements improves the accuracy of the parameter estimates by a factor of $\sqrt{(N-1)}$, but does not improve on the question whether a maximum is present or not. Kohlbeck and Scheidegger (1985) have given specific field cases of this: if three direction maxima exist at all amongst the joints at an outcrop, they appear already with 33 measurements (taking more measurements improves the accuracy of the parameters only slighty); if there are no maxima present, even hundreds of measurements won't produce any. Inasmuch as for the comparison of joint orientations with other morphometric data, only the steeply dipping joints are recorded, only 21 measurements at each outcrop are sufficient. According to the usual rules of measurement theory, three *outcrops* (i. e. 63 joint orientations) are needed to establish "regional" orientation maxima.

The practise of taking three measurements to support a mean can be supported by statistical reasoning. The 95% or 99% confidence limits l of a mean of N measurements depend on a "Student-function" $t(N)$ in the following manner

$$l = t(N) \cdot s/\sqrt{(N-1)} \tag{1.17}$$

(cf. any textbook on statistics, e. g. Marsal 1967) where s is the root-mean-square deviation of the sampled data from the sample-mean. The function $t(N)$ drops from very high values to below 10 for $n = 3$, asymptotically approaching a value around 2 for $n > 3$. In fact, for regional results one does not average the mean values of the joints for each outcrop, or for the individual river links or ridge segments in the area, but treats all the data for joints, river links or ridge segments together. Thus, the sample sizes are generally much bigger than absolutely necessary.

1.4
Tectonic Features

1.4.1
General Remarks: The Tectonic Stress Field

Obviously, tectonic features are a significant part of *morphotectonics*. Such features are caused by a tectonic stress field.

"Stress" is physically a tensorial quantity which must be represented by six individual "components". A stress tensor has three principal axes; these are the axes on whose polar planes solely tensional or compressional forces are acting. Shearing stresses are absent. Inasmuch as the stresses vary from point to point in the Earth, they form a *stress field* in the Earth. The lines that are always tangent to a principal stress (e. g. principal pressure) direction are called its stress trajectory (e. g. pressure trajectory). Inasmuch as the surface of the Earth is a free boundary, across which no shear stresses can be transmitted, the principal stress trajectories near the Earth's surface must be close to horizontal or vertical. Thus, in morphotectonic investigations, one is essentially faced with two horizontal stress trajectories; the third principal stress trajectory is approximately vertical.

In order to measure the stress at a point, one would have to determine six components. It would be handy if an instrument would exist which, upon being stuck into the Earth, would directly indicate the values of the six required components. Since such a gadget does not exist, the most convenient way is to fix an extension strip ("door-stopper") to the end face of a drill hole, to overcore the strip and to measure the extensions undergone by the strip upon stress relief (Hast 1958). From the measured extensions, the stresses have to be inferred by making a corresponding calculation. To do this, the elastic moduli of the core, which have to be measured separately, have to be obtained. In this fashion, one acquires true "in situ stress measurements". There are other, equally tedious methods for doing this, such as observing breakouts in oil wells or stress restoration methods. World data of in-situ stress measurements have been collected, notably by Zoback (1992) on a map.

Inasmuch as in-situ measurements are rather tedious and costly, one can try to determine at least the orientations of the principal stresses from geological observations in the field. In particular, this can be done by studying the nature and orientation of joints and of faults. In addition, the investigation of earthquake source mechanisms, in particular of their fault plane solutions, can give information on the tectonic stress field. Paleostresses can be found from studies of petrofabrics, notably of stylolites.

1.4.2
Joints

1.4.2.1
Definition

Any visible small (5 cm to several meters long) crack or fissure in rock is called a "joint". Joints are ubiquitously present on rock outcrops, road cuts, mountain sides in materials ranging from firm, solid plutonic rocks to extremely friable recent sediments. It is generally surmised by the scientific community that global tectonics has something to do with their genesis (Scheidegger 1978, 1985c; Hancock and Engelder 1989, Hancock 1991). Very often, joint directions correlate with those of other geomorphological features such as rivers, gorges, fiords and ridges (Hantke and Scheidegger 1999). Such a correlation might point to a genetic relationship between the various features.

1.4.2.2
Joint Patterns and Joint Sets

Although a rock ledge may be dissected by many joints, it is often possible to recognize three sets as forming the surfaces of the fundamental rock "cell":

Fig. 1.11. A series of breakout niches on a rock ledge featuring fine-grained sandstone of the Upper Marine Molasse (Miocene) on the Ufenau, an island in Lake Zurich, Switzerland, showing two subvertical and one horizontal joint sets

evidently, three fracture sets are needed to define a block of rock (or its counterpart: a niche). Such a block is called "fundamental parallelepiped"; very often it is visible as such in outcrops (Fig. 1.11). Mostly, however, the picture is one of great complexity, but usually, three main sets are recognizable; of the three main sets, one runs generally more or less horizontally and is obviously connected with the lithology, the other two are more or less vertical and at approximately right angles ("conjugate") to each other, and are regarded as tectonically induced. Thus, exposed rock ledges often acquire a jagged appearance during disintegration, indicating three distinct joint sets, two of which are steeply dipping; their strikes are oriented more or less at right angles to each other ("subvertical" *conjugate* joint sets); a typical example is shown in Fig. 1.11. These regularities are often not seen as clearly as in Fig. 1.11, but the pattern has been identified as frequent throughout the world by the statistical evaluation (cf. Sect. 1.3) of joint orientation measurements in many thousands of outcrops on all six continents. In fact, it is a rare occurrence indeed that the indicated statistical methods do *not* identify the three basic joint sets. The author knows of only one case (Hallstatt, cf. Kohlbeck and Scheidegger 1985) where complete chaos was found; in about 5% of the outcrops only *one* axis (in addition to that representing the subhorizontal joint set) of the orientation tensor was identified, in the rest all three (i.e. two subvertical joint planes) axes were recognizable.

1.4.2.3
The Age of Joints

Since joints are surface phenomena (unless they are manifestly filled with some "old" material), they cannot be older than the surface itself. Engelder (1985) noted that joints propagated only to a depth of 500 m because they are absent in cores taken at greater depth, an observation that the present author also made in deep mines. In tectonically active regions, such as the Himalaya (Kizaki 1994), uplift and denudation rates are of the order of up to 10 mm/a, i.e. 10 km/Ma – thus what one sees at the surface, including the joints, must be all very young (possibly less than a few tens of thousands of years old).

1.4.2.4
Origin of Joints

The problem   The "nature" of joints has generally been studied on a local basis: marks on the joint surfaces have been investigated considering indications as to their genesis. In this instance, plume structures indicating an extensional origin have been found on most joint surfaces, however, slickensides indicating a shearing movement also exist. In order to assess the significance of these observations with regard to the far-field tectonic stresses, an *extrapolation* of the local genesis of joints to far-field conditions is required. However, the scaling of small-scale features to conditions of plate tectonic dimensions is always problematical: the mechanisms that result in jointing are quite diverse (Price 1966, p. 127) so that applications of general fracture principles have to be pursued with caution (Bahat, 1991, p. 58).

Extensional origin of joints   Fractographic studies show that joint surfaces often present a characteristic plumose texture (Pollard and Aydin, 1988; Bahat 1991) which identifies them as extension fractures; joints exhibiting shear marks (e.g. slickensides) at their surfaces *do* exist, but are rather rare. If extension (i.e. pressure relief $T = \sigma_3$) is responsible for the formation of most joints, the joint planes are parallel to the maximum compressional stress ($P = \sigma_1$), prevailing in the vicinity and dilatation-extension determines joint initiation and propagation (Price 1966). A simple extrapolation of the fractographic evidence to far-field conditions would indicate that joints are indicative of the greatest/smallest-pressure direction in the *large-scale* neotectonic stress field. Unfortunately, tensional fractures can never easily produce conjugate joint sets on a large scale, since only *one* maximum compression direction exists at any given location in a stress field. However, recently Caputo (1995) proposed a mechanism featuring re-orientations ("swaps") of the local principal stresses during formation-"cycles" of the joints: In surface rock subject to a uniform biaxial horizontal stress, fractures occur at right angles to the least principal stress (T) direction when the tensile strength is locally exceeded. Then, Caputo proposes that this direction experiences a stress drop due to the stress release; the stress field, retaining the same principal directions, is thereby locally distorted by a "swap" between T and P. When failure conditions are again

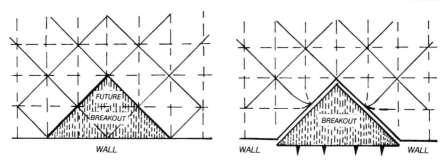

Fig. 1.12. Schematic drawing (in plan) of a vertical ledge with a breakout niche (breakout material *stippled*) and stress conditions (*solid lines:* shear lines; *broken lines:* principal stress trajectories) in its vicinity: *left picture* shows the conditions before the breakout, the *right picture* shows the deflection of the stress trajectories and shear lines during/after the breakout (modified after Scheidegger, 2001a)

reached, further fractures form at right angles to the previous ones. Repeated cycles of failure-events, stress drops and stress swaps eventually generate a "conjugate" subvertical joint system. Such a mechanism could indeed produce fracture gridlock systems on large flat surfaces.

Shearing mechanisms   "Conjugate" fracture-patterns are common in all kinds of materials; they are caused by the shear in the applied stress field: Shear lines (at the surface) occur in conjugate pairs and their *bisectrices* are the directions of the principal directions of the stresses that caused them. Thus, the morphological appearance of joint *sets* (in contrast to that of the joint *surfaces*) would favor a view that the latter are shearing features of the large-scale stress field that caused them. Although contradicted by the fractographic evidence, a *shearing origin* of joints would have the advantage that the joints would automatically occur as a grid of "conjugate" sets without a complicated stress-swap mechanism: in any stress field, there would be (in plan) two conjugate shear directions at right angles to each other. This fits the observation that subvertical joint sets occur frequently (at least statistically) in conjugate systems. This does not mean that joints would necessarily be shearing *fractures*: in the latter the angle between fracture planes would have to be substantially less than 90° (Mohr [1928] suggests about 60°), but in natural joint systems the angle between the conjugate joint planes is mostly close to 90°; thus the author suggested long ago (Scheidegger 1978) that the joints may not actually be shearing *fractures*, but some other type of shearing *phenomena* like Lueder's lines in a ductile material or a surface-response to the shearing strain in the underlying tectonic plate. In either case, the stress trajectories of the tectonic stress field would be in the direction of the bisectrices of the strikes of the local conjugate subvertical joint sets.

Reconciliation of mechanisms   Therefore, the relationship between surface joints and the tectonic stress field still constitutes, in spite of the assertion of many geologists that joints are extension features in large-scale tectonic stress fields, a problem (Pollard and Aydin 1988) and a *reconciliation of the conflicting observations* is required. Scheidegger (2001a) has suggested that the plumose texture of many surface joint planes could indeed be locally due to stress relief and not directly due to the far-field stresses; during the immediate disintegration of a rock face there is obviously an extension present at the fracture surfaces during the last moments, otherwise the material could not separate out of the rock body. The classic example of this circumstance represented by the breakouts at the rock ledge shown in Fig. 1.11 could be explained by the new mechanism proposed in Fig. 1.12: The far-field principal stress trajectories *must* be parallel and normal to the overall trend of the ledge on account of the boundary conditions: air cannot support a shear (Fig. 1.12a); the "niches" break out in wedges along the shear lines, but the local boundary conditions now require a deflection of the principal stress trajectories to be at right angles to the developing breakout surfaces, causing tensional stress relief on them (Fig. 1.12b). It is therefore possible that continental scale *conjugate* joint sets are *designed* by shear in the surrounding neotectonic stress field, although individual joints may feature extension markings caused during the last moments of rock disintegration.

Evaluation: Joints and geomorphotectonics For many morphotectonic studies, the problem of whether joints are shearing or extension features in the surrounding tectonic stress field is of no importance. Thus, observations such that joints form subparallel conjugate sets over whole continents (Scheidegger 1995) or that morphological features (see Sect. 1.5), e.g. drainage patterns [Prasad 1979; Eyles et al., 1997] and lake shore orientation [Eyles and Scheidegger 1999]) are often parallel to the joint orientations are independent of the mode of formation of the joints. The differences between the two models arise only in subsequent interpretations: As noted above, if the joints are designed as *tension/pressure* features by the far-field stresses, then the principal stress directions are in the direction of the *strikes* of the subvertical joint sets; if joints are designed by *shear* in the surrounding tectonic stress field, the principal stress directions are in the direction of the *bisectrices* of the subvertical jointsets; the two directions differ by 45°. Evidence of the frequent occurrence of the shear design of joints (and therewith of many geomorphological features) is widely available and will be presented in Chap. 2 (Global Morphotectonics). It is also worthy to note that the bisectrix of the *smaller* angle between the joint sets indicates in most cases the maximum pressure (P) direction of the stress field derived by other means; this would correspond to the Mohr criterion for shearing fractures. In the tables, we shall always list this bisectrix of the smaller angle as the *first* bisectrix.

1.4.3
Faults and Lineaments

1.4.3.1
Faults

Faults are the "large editions" of joints. Faults are discontinuous fractures within the Earth which may reach to the surface. A fault, like a joint, is characterized by the plane (numerically dip direction and dip angle) in which the motion occurs (the "fault plane") and the by the direction of motion thereon ("slip angle", angle between the motion vector and the "strike" of the fault; the latter is the intersection line between the fault plane and the Earth's surface). The motion direction is also often characterized by the plane to which the motion direction indicates the polar axis (auxiliary plane). Fault- and auxiliary planes are always at right angles to each other, they are "conjugate". For geological purposes, the fault strike is a more preferred characteristic. In the geological nomenclature, it is usual to call a fault "transcurrent" (or "strike-slip") if the slip angle is close to zero. If the slip angle is close to 90°, the fault is called "normal" if it induces an extension of the surface, otherwise "reverse" (Anderson 1951). Sets of faults can be represented by pole density diagrams or by strike roses. The statistics of faults follows the principles of general orientation statistics of spatial axes as outlined in Sect. 1.3.

Faults have a direct connection to tectonics, inasmuch as they are the result of failure of the uppermost crust of the Earth under the action of tectonic stresses. As noted above (Sect. 1.4.1), because the Earth's surface is a free boundary, the principal stress trajectories near it must be approximately horizontal or vertical; this produces essentially the three types of possible faults (transcurrent, normal and reverse) as mentioned above. Normal faults produce fault scarps at the Earth's surface and are thus the most easily recognizable; therefore they are perceived to be the "normal" type, although they are by no means the most frequent. The transcurrent faults are of greatest tectonic interest, because they are indicative of a stress field in which the maximum and minimum pressures are horizontal. These faults are steeply dipping, their strikes are more or less at right angles conjugate to each other. According to the Mohr (1928) criterion, the maximal and minimal pressure are the *bisectrices* of conjugate fault surfaces (cf. the analogous situation with joints). These may be the present-day neotectonic stresses and the fault may still be in motion occasionally like the San Andreas fault in California, or they may be "fossil faults" like the Diendorf fault in Lower Austria.

1.4.3.2
Lineaments

Lineaments are (possibly winding) linear features on the Earth's surface that are identifiable on maps, aerial photographs or satellite imagery. Hobbs (1912) defined them as "significant lines of the landscape which reveal the hidden architecture of the rock basement". The orientations of lineaments can be treated statistically by the methods outlined in Sect. 1.3, if necessary after sectioning them into segments that can be approximated by straight lines.

However, although the lines are easily observed on the images, their nature and significance are not easily assessed; in most cases, field inspection on the ground is absolutely necessary.

Thus, in the author's experience, a very definite linear feature (*Sandberg Lineament*), visible on air photographs of the region around Hollabrunn, 50 km NW of Vienna (Austria) turned out to be the outcrop of a gently SE dipping sand horizon with no relation to marked faults in the crystalline basement (Figdor et al., 1983).

Similarly, the basic genesis of the *Niagara Escarpment*, running from Hamilton to the northern tip of the Bruce Peninsula and on to Manitoulin is usually assumed as inherent in the cuesta-type of landscape characteristic of SW-Ontario (Lattman 1968): The bedrock of south-western Ontario consists at the Lake Erie level of SW-dipping (ca. 7°) Silurian strata covered by Late Pleistocene glacial sediments in which differential erosion occurs. This is underlain at the Lake Ontario level by Ordovician shales. The Niagara Escarpment is an exposed erosional scarp in the Silurian strata between the Lake Erie and Lake Ontario levels: its rim is formed by Middle Silurian dolostones (Lockport-Amabel Formation; Johnson et al. 1992); the latter are underlain by clays of the lowermost Silurian ("Cataract") formation which are eroded rapidly, giving rise to the escarpment reaching over hundreds of kilometers. Nevertheless, on geological maps, the Niagara Escarpment appears as a "lineament" that runs for most of its course almost in a straight line oriented N165 E. All the major reentrant valleys cutting back into the Escarpment are obviously at right angles to this. Thus, there must be a significant tectonic factor present in its genesis.

1.4.4
Earthquakes

Earthquakes can have a direct effect on the morphology, such as when active faults show displacements on the Earth's surface. The studies of source mechanisms are however of greater interest in connection with morphotectonics, since the latter can give an indication of the nature and orientation of the neotectonic crustal stress field.

In a crude approximation, the phenomena occurring at the source (focus) of an earthquake can be considered as a fracture type process: It is as if a spherical region around the focus ("focal sphere") is cut in half along a fracture plane (cf. Fig. 1.13) along which the upper part moves one way, the lower part moves the opposite way. For such a displacement to take place, one has to assume a stress field in which the maximum (P or σ_3) and minimum (T or σ_1) pressure exactly act in the quadrants which are formed by the fracture (or "fault"-) plane and the plane normal to the displacement direction ("auxiliary plane"). Thus, the principal stress directions can be ascertained for every earthquake of which there are sufficient records to make a fault plane solution.

The above mentioned fracture model works phemenologically very well for earthquake foci, although it is difficult to envisage brittle fracture as being

Fig. 1.13. Model of an earthquake source

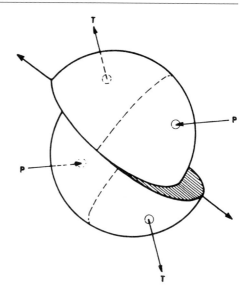

able to take place in intermediate and deep-focus earthquakes (Scheidegger, 1982a). Moreover, the stress equilibrium during the shock formation is not preserved during the fracturing process. However, the phenomenological appearance of earthquakes can still be easily visualized as that of a shock from a brittle fracture. Proper mechanical models envisage double dipole sources and seismic moment tensors at the source (cf. Aki and Richards 2002).

Nevertheless the fault planes can be determined from the analysis of the first motions (push or pull) along the rays that leave the focus at various directions, a method originally initiated by Byerly (1938) in connection with a study of an earthquake in Northern California. As for any other fault plane, the orientation of an earthquake fault plane in space can be numerically defined by giving its dip direction and dip and the slip angle of the motion vector; the latter can also be defined by its polar ("auxiliary") plane (see above). Fault and auxiliary planes are at right angles to each other ("conjugate"), the principal stresses are the bisectrices of fault and auxiliary planes. The statistics of sets of fault planes follows the same scheme as general orientation statistics (cf. Sect. 1.3).

The number of actually available fault plane solutions is legion. Inasmuch as the source mechanism is a manifestation of the neotectonic stress field, the latter can be reconstructed from a series of fault plane solutions of earthquakes in an area.

1.4.5
Petrofabrics

Finally, the tectonic stresses can also affect the "fabric" of rocks: small components of the rocks may be affected or even created by the tectonic stress field. Thus, pebbles in conglomerates may become deformed and crystals may be squeezed by stresses. Inasmuch as such effects can be retained over millions of years, they may be indicative of "paleostresses" that had been active long before the present. By dating the formation time of the stylolites, paleostress fields can be reconstructed for various geological times.

The most interesting petrofabric features are *stylolites*: they are needle-like pressure-solution phenomena which occur mainly in limestone and cherts: precipitates grow from a given surface in the direction of the largest pressure present during their formation (Schaefer 1978). Horizontal stylolites are particularly interesting because they indicate the direction of the surface trajectories of the largest pressure that was present at the time of their formation.

1.5
Planar Morphological Features

1.5.1
General Remarks

Many morphological features are characterized by a series of lines winding on a two-dimensional surface, the latter usually represented by a planar map. The lines themselves as well as sets of such lines have orientation structures which can be analysed by the methods discussed in Sect. 1.3. However, these methods are designed for making statistical analyses of *linear* elements, so that it is necessary, first of all, to "rectify" the lines or sets of lines, in order to obtain sets of numbers representing azimuths. The ways of doing this can be different for individual "wiggly" lines or for networks.

1.5.2
Wiggly Lines

Typically, streams, ridges, fault scarps and coasts (also lake shore orientations) are represented on maps by winding lines. In fact, such lines are not one-dimensional, but have fractal dimensions between 1 and 2. A (linear) "length" cannot be defined for them as shown by Mandelbrot (1967) e.g. for the coast of Britain. Therefore, the lines should not be called "winding": they are, in a true sense, *wiggly*: The closer one looks, i.e. the larger the map scale, the more wiggles appear. Truly fractal features are self-similar relative to the scale (at least within a certain range); their basic structures (such as the orientation structure) are scale-invariant (Turcotte 1992).

For an analysis of the orientation structure of wiggly lines, the latter are simply segmentized: A divider of a given constant length (say 200 m, 500 m or 1–2 km in natura) is walked along the "blue line" representing a stream on a map (or along a crest line or whatever may be of interest) and the directions (azimuths N > E) of each "segment" resulting by joining the sequential divider-step points by straight lines are noted. In this procedure, the length of the step (it must be constant, though) and the map scale do not matter, as long as the self-similarity range is not exceeded. Thus, a set of numbers (azimuths) is generated which can be analysed by the statistical methods outlined in

Sect. 1.3: the segment directions are equivalent to the strikes of joints (the polars are equivalent to dip directions), the "dips" are simply 90°. In fact, more efficient programs are also available which assume *a priory* circular rather than spherical input data (Kohlbeck and Scheidegger 1985).

1.5.3
Networks

In principle, the digitization of *network* orientations can proceed in the same fashion as that of single wiggly lines: every link in the network is segmentized by the divider-step method. This is, indeed, what is usually done.

However, a short-cut useful particularly for drainage networks exists: The drainage network is considered as a graph, specifically as a bifurcating arborescence. Each end-point (free vertex) and each confluence-point (inner vertex) of the graph is marked metrically correctly on a map; then the wiggly stream-lines (links) are replaced by *straight* graph-"edges" between the vertices (Scheidegger 1979) whose orientations (azimuths) and lengths (statistical weights) can be accurately measured. These are then used for a two-dimensional statistical analysis. This "rectification" of a stream network may appear as a somewhat "brutal" procedure, but it is certainly statistically correct; it depends only on the "blue lines" on a map and not on any interpretation by the researcher. Because of the scale-invariance of maps, it also does not depend on the map scale and is only slightly less accurate than segmentizing each stream link individually, because of the fewer number of input data in a particular river network. However, in view of the greatly increased power of computers since 1979, the present practice is individual segmentization.

1.6
Significance of Correlations

1.6.1
Local Correlations

First of all, it is of interest to compare the various morphometric observations within one region with each other: How do the river or ridge patterns relate to the joint patterns? Quite generally, if there is a direct correlation, a common genetic origin, e.g. by the neotectonic stress field, can usually be surmised. Furthermore, it is generally desirable to relate the local morphometric observations to other *physically identifable features*, such as the mid-Atlantic ridge with offsets, extended fault lines, volcanic plateaus and the positions of the foci of earthquakes.

For such a type of studies, it is immaterial whether the joints indicate the shearing or principal stress directions of the local neotectonic stress field. If comparisons with other types of stress determinations (from in-situ measurements, fault plane solutions of earthquakes or stylolite orientations) are made, this does matter, of course. The general observation is that joints usually indicate the shearing rather than the principal stress direction in most cases.

1.6.2
Plate Tectonics

A further, much less unambiguous undertaking, is the *interpretation* of the morphometric facts in terms of some global model, such as global "plate tectonics". The latter was developed in the second half of the last century (Dietz 1961, Hess 1962) based upon earlier ideas of Wegener (1915). Briefly, it assumes that the Earth's lithosphere is divided into a relatively small number of plates (Fig. 1.14) at whose boundaries much of the tectonic, seismic and volcanic activity occurs; the plates undergo displacements with each other which lead to tension zones at their margins, over- and underthrusts as well as to transform-faults (Wilson 1965), the latter causing offsets on mid-ocean ridges. Hot material rises from the mantle at the tension zones existing in mid-oceanic ridges and then spreads towards the adjacent continental margins whilst cooling, where it is subducted to compensate for the spreading. Intraplate volcanism can also occur, when a plate moves over a stationary hot spot in the mantle.

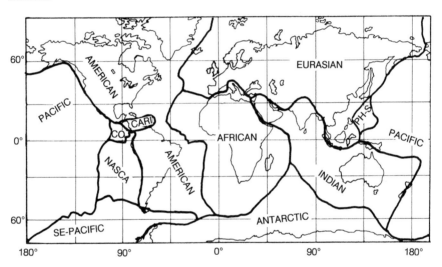

Fig. 1.14. System of tectonic plates on the globe. Names self-evident except: *CARI* = Caribbean Plate; *CO* = Cocos Plate; *PH-S* = Philippine Sea Plate

1.6.3
Other Models

However, difficulties in the above-mentioned idealistic scheme have become more and more apparent lately. A comprehensive collection of "problems with plate tectonics" has been published by Augustithis (1990); more recent criticisms were presented by Meyerhoff et al. (1992b), Shields (1997), Storetvedt (1997, 2003), Wright (2000) and others. Thus, new reflection studies seem to suggest that the mid-Atlantic ridge is a gigantic bivergent structural fan that originated in much the same way as continental foldbelts (Meyerhoff et al. 1992a). The plate boundaries between the African and Eurasian plates are uncertain: Zazo and Goy (1994) have proposed an additional "Iberian" (micro-) plate sector between the Eurasian and African plates with its apex in the Azores. The causes of segmentation of the mid-ocean ridges by long-lived transform faults is not properly understood (Bonatti 1996). The rates of plate movements between the Eurasian and American plates have traditionally been quoted (from the comparison of palaeomagnetic polarisations across the mid-Atlantic ridge and those in rock cores which can be dated) as reaching 40 mm/a (Heirtzler et al., 1968). However, Global Positioning System (GPS) measurements between Eurasia and America (Kogan et al. 2000) indicate movements of only ca. 1 mm/a. Jonsson et al. (1999) interpreted GPS measurements on S. Miguel as indicating spreading movements of only 4 mm/a between the African and Eurasian plates, but the authors themselves warn that these movements could also be the result of an inflation of the magma chambers below. Furthermore, the long-held assumption that hot spots are fed by deeply-fixed mantle plumes may not be true (Wright 2000, Pinsker 2003). The various authors mentioned usually also suggest alternative models: Meyerhoff et al. (1992a,b) magma surge channels during contraction of the Earth, Shields (1997) an expansion of the Earth, Storetvedt (1997, 2003) wrench faulting caused by periodic realignment of the equator due to redistribution of masses in the Earth's mantle by oceanization of an originally wholly continental crust. Evidently, there is not much agreement amongst the various views. Nevertheless, some sort of mobile plate structure in the Earth's lithosphere still seems to be the most commonly accepted model, especially since the various alternatives are not really entirely convincing either. We shall, therefore, aim our review mainly at the *morphometrically* establishable relationships between joints, ridges and streams, trying to fit them into the *physically* evident tectonic frameworks and attempt to discuss them only marginally within the framework of *interpretative* global tectonic models.

CHAPTER 2

Global Morphotectonics

2.1
Scope of Chapter

We are now proceeding to present a review of the work done on joints on all six continents and on islands in two oceans. This will update the author's earlier reviews of the subject (Scheidegger 1985c, 1993a, 1995) with emphasis on new material: It will be shown that a much broader synthesis can now be obtained of the subject than was possible heretofore: In practically all instances, a direct connection between orientations of joints and those of other planar geomorphological and tectonic features can be established.

Inasmuch as this chapter is concerned with global tectonics, we use the traditional (plate tectonic) division of the world into Laurasia, Gondwanaland and the oceans for convenience: Laurasia encompasses Europe, Asia (excepting Peninsular India), North America and the Arctic (being thought as having been formed at some earlier geological time from a single continent), Gondwanaland Africa, India, South America, Australasia and the Antarctic (which also have been thought to have been united at one time); as oceans we consider the Atlantic, Indian and Pacific oceans as fundamental, with the other oceanic areas as "adjacent" to their respective continents.

2.2
Laurasia

2.2.1
Europe

2.2.1.1
Morphology/Geology

Europe is essentially a peninsula attached to the west of the Eurasian land mass; it is physically separated from Africa by the Mediterranean Sea. Its ancient nucleus is the Baltic Shield, a region of rocks that has remained stable since the Precambrian. Towards the E and the S the old rocks of the Shield are covered by thin flat-lying sediments: the Russian Platform, continuing into the N German plains and the English Midlands. Subsiding intermittently, this

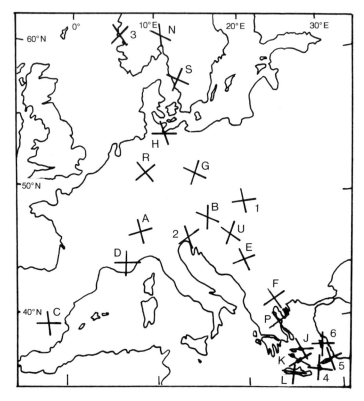

Fig. 2.1. Preferred joint strikes in Europe. Signature keyed to Table 2.1

area was flooded at times by shallow seas. The triangle formed by the Shield and Platform is bordered on its three sides by orogenic belts: on the NW by the Caledonian belt (post-Silurian orogeny) extending through Scandinavia to Britain, on the E by the Urals (Carboniferous–Permian orogeny) and on the S originally by the (also Carboniferous–Permian) Hercynian belt which was later superseded by the Alpine (and associated Jura) Ranges of post-Cretaceous folding age (Holmes, 1944). The Alpine belt is tectonically (uplift, earthquakes, volcanoes) active today; it diverges into several branches (Carpathians, Dalmatian mountains with extensions to Greece and Crete); the plate boundary between the African and Eurasian plates runs in its vicinity (Periadriatic [or Insubric] lineament). European joint orientations have been reported by the author (Scheidegger 1995). In addition, further measurements have been made in additional areas. A summary of these observations is given in Table 2.1 containing the corresponding preferred joint strikes and their bisectrices which may indicate the principal stress directions (Fig. 2.1).

Table 2.1. Strike/trend directions of joints and morphological elements in Europe. Signatures keyed to map (in Fig. 2.1)

Loc.		No.	Max 1	Max 2	Angle	Bisectrices	
Scandinavia							
3	Bergen area						
	Joints	509	147±09	47±04	80	7	97
	Fiords	51	151±06	51±11	81	11	101
	P-Earthq.	18				96	
N	S-Norway						
	Joints	209	136±10	55±09	82	95	6
	Valleys	78	140±05	(02±02)	41		
	Lineaments	73	149±01	48±07	79	8	98
S	Bohus (Southern Sweden)						
	Joints	789	107±00	23±01	84	155	65
Central Plains							
H	Kiel	5	160±07	90±33	70	124	35
R	Ruhr Area	31	140±10	50±08	90	94	4
G	Eisenach	29	117±20	19±22	82	158	68
U	Balaton	21	122±22	21±07	79	162	71
Alps-Carpathians							
D	S. Alps	402	91±07	2±08	90	137	46
A	Switzerl.						
	Joints	29*	156±04	70±08	86	113	23
	Valleys		145±02	53±02	88	99	9
	P-earthqu.	13				145	55
B	Austria						
	Joints	28*	116±22	0±20	64	148	58
	Valleys		82±03	2±03	80	132	42
1	Slovakia	167	172±01	86±00	85	129	39
Spain							
C	Spain						
	Joints	170	174±10	91±12	84	133	43
	Valleys	479	18±08	90±08	72	144	54
E-Adriatic							
2	Trieste	96	161±20	57±08	76	18	109
E	Visegrad	66	156±26	72±05	76	114	23
F	Macedonia	180	144±12	57±06	89	100	11
P	Pelion	398	157±02	71±05	86	114	24
Aegean							
J	Naxos	21	151±23	89±12	62	120	30
K	Santorini	21	120±15	64±04	56	93	2
L	Crete	232	9±00	96±08	87	52	142
4	Karpathos	331	7±03	97±00	89	143	53
5	Rhodos	405	168±06	77±00	90	32	122
6	Kos	173	8±08	90±06	82	49	139

*regions containing some 60 to 2000 individual measurements

2.2.1.2
Scandinavia

Field observations show that the surface joint systems in Sweden correlate with those in central Europe; they are the result of the intracratonic stress field and the mechanical response associated with the Alpine orogeny. The stress systems in southern Norway, on the other hand, are the result of the ongoing extensional or wrench-fault tectonism in the Atlantic crust associated with the stresses near the mid-Atlantic Ridge, which are normal to the contiguous coastlines from Scandinavia to France, Portugal and North Africa (Zheng and Scheidegger 2000). The dominant joint systems in southern Scandinavia relate to the stresses caused by the neotectonic movements: The Alpine tectonism leading to a NNW-SSE and/or NW-SE compressional stress regime is considered to have caused the N-S and E-W joints in Sweden. The almost N-S and E-W directed principal stresses in the southern areas of Norway (latitudes $< 62°$) are related to the Atlantic tectonic regime (ridge push towards the coasts of Scandinavia, France, Portugal and North Africa) and the ensuing neotectonic intracratonic stress regime: Thus, there seems to be a fundamental tectonic dividing line along the Skagerrak, continued by the Oslo Graben. According to Storetvedt (1997, p. 290) this dividing line can even be followed as far as the Rhine Graben.

2.2.1.3
Central Plains

The Central Plains in Europe run from the Atlantic coast of Germany inter-mittently into Poland and Hungary. Measurements have been reported at a few spots in this area (Table 2.1). It is observed that in the central part of the European plate the principal compression direction is oriented NW-SE. This fact has been known for a long time; it also correlates with interpretations of visible fault structures, earthquake epicentral zones and fault plane data (Bankwitz et al. 2003).

2.2.1.4
Alps-Carpathians

Geologically, the Alpine–Dalmatian mountain ranges running from southern France through Switzerland and Austria to the Balkans are the most interesting parts of Europe. These mountain ranges are part of a global belt extending onward through Turkey, the Hindukusch and the Himalaya, into Burma and Indonesia. The tectonism occurred at and after the time boundary between the Cretaceous and the Tertiary periods and is ongoing today. Many measurements have been made along it. Furthermore, in such areas as were investigated in this regard (Switzerland, Austria), the joint strikes correlate with the preferred directions of the river segments (Scheidegger, 1979a,b). The stresses in these regions are presumably intraplate stresses in the Eurasian tectonic plate.

2.2.1.5
Spain

Spain is a somewhat distinct part of Europe, separated from its main body by the Pyrenees in the north; another mountain range, the Sierra Nevada, exists in the south. Our measurements (Scheidegger 1979c) were made on 11 outcrops stretching along the SE coast and the Sierra Nevada. Valley trends were also measured on five main rivers (Ebro, Jucar, Guadalquivir, Tajo and Duero). The trends of the rivers correlate closely with those of the joints.

2.2.1.6
East-Adriatic

The formerly Yugoslavian and the Pelion-Sporades region and south thereof have substantially different joint orientations (in a counterclockwise direction) from the rest of Central Europe. It may be suspected that these areas lie in another geotectonic province which is situated south of the Periadriatic Lineament. One sees, therefore, that the joints represent a signature of plate tectonism inasmuch as the Periadriatic Lineament is an old plate boundary between Africa and Europe, and that the stresses to the S of the Lineament represent rim stresses of the Eurasian tectonic plate (Scheidegger 1995).

2.2.1.7
The Aegean

Cyclades   The trend from the Sporades is continued in the Cyclades and Crete. The structural nucleus of the Cyclades is an ancient crustal block related to the Rhodope mountains of Greece which has been severely fragmented (Jacobshagen 1986); it is bounded in the S by a volcanic arc which contains Santorini (Thera), an island which is the remnant of a large volcano that exploded in (pre-)historic times. Basically, the trend of the principal stress directions rotate southward in a counterclockwise direction; one is nearly NS in Santorini (corresponding to a NS T-direction [i.e. a NS-extension] which has also been found by in-situ stress measurements [Paquin et al. 1984] and structural analyses [Angelier 1977]).

Crete   The above trend becomes NNE-SSW in Crete. The island is part of the 650 km long South Aegean Arc connecting the Hellenides of the Greek mainland (Peloponnesos) with the Taurides of SW Anatolia. This arc was thrown up as a mountain range during the Tertiary Alpine orogeny in four stacked S-vergent allochthonous units (nappe-piles) thrust over an autochthonous basement. Thus, on all islands of the Arc there is an autochthonous base and four allochthonous units, which, however, were often broken by fractures, rifts and dislocations and disappeared beneath the sea. In Crete, the fractures left an E-W sequence of four main mountain masses standing above the intramontane depressions which are now isthmusses composed of Pliocene and younger sediments: The whole is an E-W en echelon structure where deeper facies occur in the W than in the E. During the process of rifting, the residual mountain masses were upheaved and tilted toward the N and NE, with the result that while on the S side they fall abruptly into the Libyan Sea, they descend more gradually to the N (Jacobshagen 1986). Thus, on the S side, particularly in the W of the island, there has been a possibility for the development of some spectacular watershed gorges such as the Samaria gorge. Joint orientation measurements were performed in 9 areas. Whilst the results for the individual areas scatter considerably (with large error estimates, up to 47°), the result for the island as a whole (see Table 2.1) is very definite: the strikes are N9° $\pm$ 0°E and N96° $\pm$ 8°E. It was noted that the first of these directions is parallel to the intermontane depressions which strike NNE-SSW. The second direction is evidently conjugate thereto. Since the roads follow these depressions or run at right angles to them, the joints are parallel to the highway system of Crete. Thus, the joints have likely been created by the same forces that created the en echelon faults separating the four mountain massifs.

Karpathos Karpathos is the third largest island in the Dodecanese but has relatively few inhabitants. The only larger town is Pigadia, situated along a wide bay in the SE with a beach. The island is elongated extending about 50 km in the N-S direction; a mountain chain runs along its middle which reaches to 1213 m above sea level. The maximum E-W width of the island is 15 km. Young rifts separate the mountain chain into three massifs; thus it is very rugged (as had been already noted in antiquity). Like Crete, Karpathos consists of an autochthonos basement and four allochthonous tectonic units: the basement surfaces only at the West coast. The first allochthonous unit covers wide areas; it is characterized by lens-shaped in-sheared layers of gypsum, neritic cretaceaous limestones and eocene limestone-breccias, into which a flysch-like mass is tectonically intercalated. The mountains consist of the second allochthonous unit. The two highest allochthonous units are only sporadically visible in small central troughs within the mountains (Jacobshagen 1986). Joint orientations were measured at 16 outcrops stringed from south to north, mostly in limestone. The results of the numerical evaluations are summarized in Table 2.1. It is observed that the prevailing strike-directions of the joints run N-S and E-W and that the Karpathos-joints correlate very closely with those in Crete; in relation to Rhodos, there is some small difference.

Rhodos The oval island of Rhodos, situated approximately at 36°N latitude and 38°E longitude is the second-largest island (1400 km^2) after Crete in the southern Aegean; its length is 78 km, its largest width 38 km; it is characterized by a long mountain range whose largest summit (Ataviros peak) reaches 1215 m. Geologically, Rhodos is part of the S-Aegean Island Arc with its basement and four units. On Rhodos (Jacobshagen, 1986), the basement crops out in the Ataviros region (Akramitis Limestone, age Turonian to middle Eocene). The first allochthonous unit is represented by some flysches and turbidites, the second by Mesozoic-Tertiary shallow-water sediments, the third is seen only sporadically as a relic in some depressions (cherty limestone, upper Triassic to Liassic) and the fourth also only as a relic in grabens and en-echelon fractures. In addition, Neogene sediments are found above these units. Joint orientation measurements were made at 19 outcrops all along the island in all the mentioned geological units. The data were evaluated statistically; the results are listed in Table 2.1. It is seen that the prevailing strike directions run NNW-SSE and ENE-WSW, somewhat off and not really comparable to the results from Karpathos and Crete.

Kos   After Rhodos, Kos is the second largest island of the Dodecanese. Its location (Location 6) is shown on the map in Fig. 2.1. It lies very close to the peninsula of ancient Halikarnassos (now Turkish Bodrum) at the southwestern corner of Asia Minor as part of an offshore chain that reaches from Rhodos to Lesbos. Kos, around latitude 36°22′N and longitude 27°19′E extends for about 50 km from ENE to WSW; its maximum width is about 10 km and its area 282 km². The island consists of three regions: an abrupt ridge along the eastern half of the south coast (Mt. Dikaios, with its highest point Christos Peak, 847 m), a rugged pensinsula (Kephalos) in the west and a central plateau prolonged along the north coast to the ancient capital facing the mainland. Kos is separated from Kalymnos and the Bodrum Peninsula (Fig. 2.1) by a shallow submarine shelf; on its south side the sea floor descends rapidly into a small basin which is part of an important graben extending from the Cretan Sea (Kretikos Pelagos) to the gulf between the Bodrum and Datca peninsulas into Turkey. The occurrence of earthquakes, hot springs and volcanism in the area indicate the presence of tectonic activity in the region. The central highlands of Kos are dominated by schist and marble; Triassic to Jurassic flysch and limestones lie on top of these metamorphic rocks; they are exposed in the central and NW part of the island. Neogene sediments cover the limestones on the northern slope of Mt. Dikaios; recent volcanic rocks, mainly welded tuffs, cover the limestones on the central plateau to the west. Indeed, volcanic activity started on Kos during the Miocene period (approx. 10 Ma ago) continuing to the present day. About 3 Ma ago, two large domes of rhyolite erupted on the Kephalos Peninsula which have remained relatively uneroded to this day. (Higgins and Higgins 1996). A total of 173 joint orientation measurements were made at 8 outcrops on the island. These were evaluated according to the usual Kohlbeck-Scheidegger statsistical method. The results are shown in Table 2.1. Incidentally, there is a remarkable consistence between individual outcrops (not listed here). The joint orientations also agree well with those on Karpathos and Crete, but not with those on Rhodos. It turns out more and more that Rhodos is somewhat anomalous in the Aegean.

2.2.2
Asia without Peninsular India

2.2.2.1
Morphology/Geology

"Laurentian Asia" is the land area continuing from Europe to the East. Its surface features a pattern of recently folded and faulted mountain ranges on a background of old plateaus and basins and marginal plains (Holmes 1944,

p. 425 ff.). The neotectonic (Tertiary and younger) ranges are the highest in the world and are a continuation of the Alpine belt of Europe extending through the Tien Shan, the Himalayas, the mountains of Burma and Malaysia into the Indonesian islands; they are presently active with elevation rates of cm/a (Kizaki 1994) and, in the islands, with much volcanism and seismicity. South of the Himalayas, part of an old shield belonging to Gondwanaland is "attached"; the latter will be treated in Sect. 2.3.

2.2.2.2
Asian Mountain Chains

Turkey   Studies of joint orientations have been made in the mountains of Turkey (around Antalya and the upper Firat [Euphrates] valley (Buser, private communication). A summary has been given by Scheidegger (1995); we present the updated results in Table 2.2 and Fig. 2.2. As an interpretation, it has been noted that the measurements exhibit the continuing turning of the European maximum compression (one of the bisectrices) from Trieste through Visegrad and Macedonia eastward.

Himalayas   The trend is continued in the Himalayas to a NNE direction of the bisectrix concerned. Measurements have been made (Scheidegger 1979d) in the Western Himalaya (Kashmir–Ladakh), in the Himachal Himalaya (Sharma et al.2002), in the Garhwal Himalaya and in Nepal (Scheidegger 1999a,b). Evidently, what is seen here are the rim stresses on the Eurasian tectonic plate; one of the joint bisectrices, taken as the maximum compression in the Himalayas, fits well together with the idea of northward motion of India leading to a continental collision in this region. A comparison of the local joint strikes with the valley trends from the Western Himalaya through to Nepal yields the result that the preferred drainage directions correspond to the shear lines of the stress field calculated from the joints. The same is true for the river

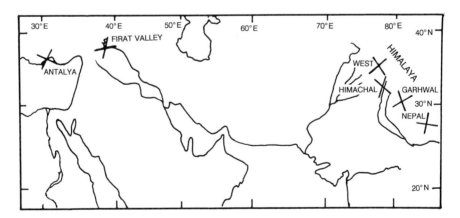

Fig. 2.2. Preferred joint strikes in Western Asia

Table 2.2. Asia without peninsular India. Strike/trend directions of joints and rivers

Loc.	No.	Max 1	Max 2	Angle	Bisectrices	
Turkey						
Antalya	44	113±17	33±16	80	73	163
Firat Vall.	764	77±17	14±03	63	73	125
Himalaya						
West-Himalaya (Jammu–Kashmir–Ladakh)						
Joints	638	145±02	48±01	82	7	97
Rivers		137±05				
Himachal-Himalaya						
Joints	218	128±12	17±11	70	154	76
Rivers	64	138±00	43±01	85	1	91
Garhwal Himalaya						
Joints	149	162±10	63±10	81	22	112
Rivers	101		54±19			
Nepal						
Joints	428	99±03	11±00	88	55	145
Rivers	178	92±03	12±05	80	52	142
China						
China West	408	130±29	47±08	86	83	5
China 104°E	614	82±21	184±16	78	43	133
China East	113	2±09	90±14	88	46	136
China South	144	174±10	80±09	86	37	127
Java	150	120±10	197±10	77	158	68
Philippines						
Joints	716	14±13	102±08	87	58	148
Rivers	999	14±04	87±07	73	52	142

directions in Tibet (Ai and Scheidegger 1981). There is an indication, thus, that the Himalayan river valleys have been designed by tectonic and not by epigenetic processes.

2.2.2.3
China

China is principally a part of the Eurasian tectonic plate with the margins towards the Indian and Pacific plates showing singular behavior. The area of North China is tectonically very complicated. A NS tectonic-seismic zone at roughly 104°E longitude separates it into two entirely different parts: the West (including Qing-Hai and Tibet, a high plateau with ongoing uplift and accretionary fold belts at its margins), and the East, essentially a block-faulted region (Precambrian cratons, in part covered by sediments) reaching to the coast. The "transition zone" in between is a "suture" (Zhang et al. 1984). To

Fig. 2.3. Preferred joint strikes
in Eastern Asia

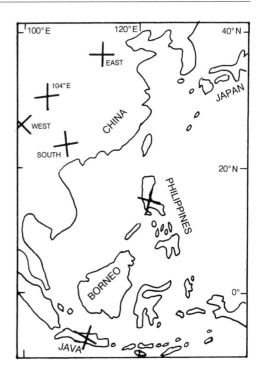

this must be added the "South" (including Hong Kong and Macao) where there
are again accretionary fold belts. Thus, one can speak of four characteristic
areas of China: "China West, China-104°, China East, and China South".

In all these areas joint orientations have been measured and evaluated (Ai
and Scheidegger 1984a). The results are given in Table 2.2 and Fig. 2.3. One
sees that there is a general push from the NE originating from the motion of
the Pacific tectonic plate. The bisectrix concerned is obviously directed more
or less normally to the strike of the mountains, which turns from an essentially
EW strike N of India (Ladakh and Tibet) to a NS direction in Qing-Hai. The
104°E zone is a singularity. These findings are supported by seismological
(fault plane solutions) and geomorphological (valley orientation) studies (see
Ai and Scheidegger 1984a). This fits well into the general picture deduced from
plate tectonics in the area.

2.2.2.4
Java

The Indonesian island arc is situated at the boundary between the Australasian
and the Eurasian tectonic plates which is a subducting plate margin. Recent
volcanic activity on this arc includes the explosion of Krakatoa in 1883 and
eruptions of many other volcanoes along the Indonesian island chain. The
explosive nature of the eruptions is due to the high gas content of the viscous

siliceous lavas involved ("Merapian" type volcanism). Joint orientation measurements were made on the flanks and in the forefield of the Merapi volcano near Yogyakarta. The top of the Merapi is covered by older Quaternary volcanic deposits, consisting of breccia, agglomerate and lava flows, including andesite and basalt with olivine. The surrounding country rocks are young volcanic deposits, undifferentiated tuff ash and lava flows. The joint measurements were generally taken in lava cliff faces on the flanks of the mountain; in the forefield the outcrops showed interbedded tuff-breccia, dacite and andesite.

Joint orientations were measured around Yogyakarta, particulary around the Merapi volcano; they were quite consistent with each other; the bisectrices were oriented N22°W (see Table 2.2 and Fig. 2.3). This direction must be considered as the neotectonic plate stress direction of the area inasmuch as it correlates with most of the fault plane solutions and the stress calculations for the area (Ghose et al. 1990). Thus, the joints on the Merapi, and presumably on the other Indonesian volcanoes as well, are caused by the plate tectonic stress field and are not cooling cracks or such like (Scheidegger 1995).

2.2.2.5
Philippines

The tectonic setting of the Philippines (cf. Fernandez 1981), is mainly influenced by the position of the region just N of the junction of three tectonic plates of the lithosphere: the Eurasian Plate, the Pacific Plate and the Philippine Plate (cf. Fig. 1.14). The present-day plate dynamics is supposedly determined by crustal underthrusting along several subduction systems and partly by strike-slip movements along transcurrent faults. The underthrusting could occur in a direction more or less normal to the strike (N145°E) of the Philippine archipelago; the normal direction thereto is roughly N145E.

Joint orientation measurements have been reported by the author (Scheidegger 1993a, 1995) from Luzon; the results of the evaluation according to the usual Kohlbeck–Scheidegger (1977) method are summarized in Table 2.2. The interpretation of this result is that one of the bisectrices (N58°E) obtained from the joints corresponds to the supposed plate tectonic underthrusting direction (N55°E) as explained above. This interpretation is supported by a statistical study of the orientation pattern of the segments of the Paygwan drainage basin: one obtains strike maxima of the gulleys at N087±07E and N014±04E, giving bisectrices at N52E and N142E. This all fits together very well with the joint orientations and with our interpretation thereof: there is a correspondence between river segment and joint orientations and both could be the result of a plate tectonic subduction from the NE.

2.2.3
North America

2.2.3.1
General Morphology/Geology

America, actually consists of two distinct land masses (North and South America) connected by a narrow land bridge. The two land masses have quite different geological histories, the northern one having developed from Laurasia, the southern one from Gondwanaland (cf. Sect. 2.3), they will be discussed separately. Morphologically, though, their structure is superficially similar: In the west there is a large chain of high recent mountains bordering the Pacific Ocean: in North America the Rocky Mountains in several major chains. To these abut in the east regions of plains, the Great Plains in North America consisting of sedimentary areas surrounding and covering a Precambrian shield. In North America the shield is bounded in the E by another, smaller mountain range (the Apallachians) which peters out into various islands.

2.2.3.2
Newfoundland

Beginning with far eastern North America we find a large island in the Atlantic Ocean which is adjacent to North America: Newfoundland. Morphologically, it consists of the Highlands, peneplained remnants of Precambrian and Paleozoic rocks. To the SE the same type of rock is tilted down to form the Atlantic Uplands. The Central Lowland is rolling country on Paleozoic rocks, largely covered by Pleistocene till and related deposits. It is unified with the mainland Atlantic provinces of Canada by a common geological history since late Paleozoic time (Poole, 1972).

In Newfoundland, a large number of joint orientations were measured and contributed to this study by Prof. K. Storetvedt of Bergen University (personal communication); we summarize here the evaluation of his data in Table 2.3. It will be observed that the orientations of these joints fit very well indeed with the those taken on the Canadian mainland, which testifies further to the fact that Newfoundland is nothing but a geological–geotectonic extension of the mainland. The plots will be found in Fig. 2.4. Furthermore, 223 river links were also investigated regarding their orientations. Only one maximum could be determined (see Table 2.3) which correlates with one of the joint strike directions.

2.2.3.3
Eastern North America

On the Atlantic Seaboard, some measurements are available from Metropolitan New York (Scheidegger 1985c). In the Canadian Shield, joint orientations have been measured in the Laurentides (unpublished measurements from Mont

Table **2.3.** Joint and river orientations in North America: strike/trend directions

Loc.	No.	Max 1	Max 2	Angle	Bisectrices	
Eastern North America						
a. Newfoundland						
Joints	1733	110±01	25±00	86	68	158
Links	221		39±01			
b. Laurentides	65	172±00	80±09	88	36	126
c. C-Ont. Shield						
Joints	5*	169±06	86±30	82	38	127
Rivers	4*	166±20	74±20	88	30	120
d. New York Cty	63	114±10	26±10	89	70	160
e. N of Lake Ont.						
Joints	5*	173±27	79±12	86	126	36
Rivers	5*	181±34	78±25	77	129	39
Bedr. chann.	5*	151±23	55±20	84	103	13
f. N of Lake Erie						
Joints	1968	164±00	104±00	60	134	45
Rivers	1950	52±00	104±00	52	168	78
Bedr. chann.	341	153±04	22±02	66	145	55
g. Pennsylvania	84	166±03	80±08	86	123	33
h. Finger Lakes	42	160±07	85±06	74	123	33
i. N New York	144	164±00	74±02	90	119	29
j. Kentucky	39	144±17	41±21	77	4	92
Western North America						
k. E-Manitoba	7962	14±04	88±07	74	51	141
l. Alberta Great Plains and Foothills						
Joints	10*	147±15	56±12	89	102	12
Rivers	12*	118±24	34±20	84	166	76
Lineaments	6*	143±10	53±05	89	188	8
m. Canadian Rockies						
Joints	25*	152±11	61±12	89	16	106
Rivers	97	150±09	43±12	73	6	96
n. Colo.Rockies	123	105±10	4±12	80	144	55
o. S-California	232	179±00	84±00	84	41	131
Mexico						
p. Center	194	170±09	78±09	88	34	124
q. South	506	129±07	38±06	88	174	84
r. Yucatán	34	116±21	48±28	69	82	172
Adjacent Islands						
t. Bahamas						
New Provid.	262	136±07	46±07	89	1	91
Bimini	105	105±10	10±11	85	148	58
Eleuthera	109	91±22	5±18	86	48	138
Gr.Bahama	154	102±13	8±13	86	145	55
All Bahamas	630	128±12	35±12	87	172	82

*Regions containing several tens to hundreds of outcrops each

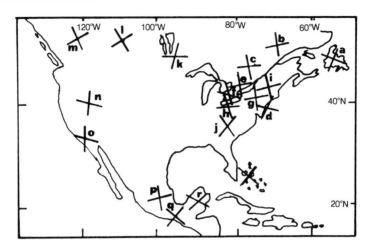

Fig. 2.4. Joint strikes in the investigated regions of North America. Letters keyed to Table 2.3

Tremblant Provincial Park) of Québec, in Northern and in Central Ontario (Scheidegger 1985c). The results of the evaluations are shown in Table 2.3 and Fig. 2.4. Measurements in sedimentary areas of E North America have been reported for the areas N of Lake Ontario (Eyles and Scheidegger 1995) and Lake Erie. The latter two studies were made in order to investigate the correlations between recent drainage channels, buried subdrift channels and joint orientations: The result was that the orientation structure of all three mentioned features in Southern Ontario are well-nigh identical so that a similar neo-plate-tectonic origin can be presumed for all of them (Table 2.3). Other studies of the joints around the Finger Lakes S of Lake Erie have been made by the author (Scheidegger 1995) and, in quite a different context, by Bahat (1993) in Northern New York: Nevertheless, the data could be processed by the statistical procedure of Kohlbeck and Scheidegger (1977) and yielded results very close to those obtained by the author in neighboring areas (see Table 2.3). Further joint orientation measurements behind the Apallachians have been made in Kentucky (Scheidegger 1985c). For all the areas of E North America discussed thus far one bisectrix seems to run NW-SE, the other NE-SW, the latter generally being identified in the northern parts of the continent as maximum pressure.

2.2.3.4
Western North America

Canadian Great Plains   Similar orientations of joints and stresses (one principal stress about NE-SW) were obtained (Scheidegger 1985c) through the Great Plains of Manitoba and Alberta as far as the Rocky Mountain Front (cf. Table 2.3 and Fig. 2.4). In Manitoba, this is confirmed by the identification and evaluation of about 8000 joints in the Rice Lake area (Scheidegger and Turek 1978); in

Alberta, thousands of linears (joint traces, river trends and photolinears) have been identified similarly, divided into 6–10 "regions" and statistically evaluated (Scheidegger 1983b). The result is again that one of the bisectrices is in the NE-quadrant. The lineament directions correlate closely with the joint strike orientations, but the river orientations do not: they are probably controlled by the general slope of the area towards Hudson's Bay, and, the region being very flat, by random meandering.

Canadian Rockies   In the Canadian Rockies, joints have been measured west of the Rocky Mountain Front in the area between Kananaskis, Calgary, Jasper, Bella Coola and Vancouver (Scheidegger 1981). The geology of this region runs the gamut from various sediments to batholiths. In the interior of British Columbia, along the Bella Coola Road, Tertiary lava and ash regions were encountered. As usual in joint orientation studies, no correlation between joints and lithology could be established. Statistically (cf. Table 2.3 and Fig. 2.4), one of the bisectrices, presumably the largest compression, is directed from the NE (N16E), like in the rest of northern North America. Furthermore, the river directions in the Canadian Rockies show a fair correspondence with the joint strikes. Thus, in contrast to the situation in the flat Albertan plains, here the rivers are tectonically controlled like the joints. Thus, one can state that one of the principal general neotectonic stresses (presumably the pressure) in Central and Western Canada is acting from the NE, a fact which is confirmed by well-breakouts and in-situ measurements (Bell and Gough 1979). The stress directions deduced from the joints are therefore confirmed by entirely different types of measurements.

Colorado Rockies   Joint orientation measurements have been made during excursions into the region West of Denver–Boulder. Denver itself is located upon more than 3600 m of sedimentary rocks which come to the surface rising steeply towards the front of the Rocky Mountains. The latter contain Precambrian nuclei (early Proteozoic biotite gneiss intruded by granitic batholiths dated from 1 to 1.7 Ga) which have been upwarped in broad anticlines during the Paleozoic and Mesozoic times. The complex rose further during the late Cretaceous (65 Ma before Present; "Laramide Orogeny") along new structural lines (e. g. Chronic 1980). The locations for joint orientation measurements, between Estes Park and the Rocky Mountain Park Visitor Center, were all situated in the crystalline Precambrian batholith. From the statistical evaluations (Table 2.3 and Fig. 2.4) one notices that the two joint strike maxima lie at N105°E and N04°E: This is more or less NS and EW, in conformity with the usual situation in North America (see Scheidegger, 1995).

Southern California   The California Division of Mines has officially divided the state into a series of geomorphic provinces, based on the shape of the land and on rock types (Schoenherr, 1992). Of these regions, the Sierra Nevada, the Coast Ranges and the Transverse Ranges are of importance to our review. The *Sierra Nevada* is an uplifted granitic batholith (peaks > 4000 m high (Mt. Whitney 4418 m); the formation and emplacement of the rock occurred during a period of 100 Ma during the Mesozoic and ended ca. 65 Ma ago, but the uplift began only about 80 Ma ago, most of it took place during the last 3 Ma and is still continuing. Recent basaltic volcanism occurred in the Sierra Nevada at its E-flank (cf. Mono Lake, Long Lake, Mammoth Mountain etc). The *Coast Ranges* extend over 800 km from near the Oregon border to S of the Santa Barbara area where they meet the Transverse Ranges at the Santa Ynez Mountains. They rise abruptly from the sea to an altitude of ca. 2000 m. They are composed of a series of NW-SE trending ridges and dip on the E beneath the sediments of the Central Valley. The rocks are extremely diverse, but, for the most part (particularly in the N), consist of oceanic sediments in various stages of alteration (Franciscan Formation). The *Transverse Ranges* lie, unlike most ranges in North America, on an EW axis. Uplift of these ranges has taken place along a series of faults that are part of the San Andreas system. Similar to the Sierra Nevada and Coast Range, most of the uplift has taken place during the last 3 Ma. The Transverse Ranges have a granitic core that is part of the S Californian batholith; it was formed in the Mesozoic age (Norris and Webb 1990, Schoenherr 1992, Dickinson et al., 1996). Joint orientations were measured in 4 areas, viz. near Thousand Oaks, in the Santa Monica Mountains S of Thousand Oaks, in the Sierra Madre and on the coast near Santa Barbara. The results of the statistical evaluation of the joint orientation measurements made, are listed in Table 2.3 and in Fig. 2.4. As is seen, one of the joint strike orientations is close to NS, the other close to EW, corresponding to the usual situation in North America.

2.2.3.5
Mexico

Regarding Mexico, we note that measurements are available from the Center (Valle de Mexico and Volcanic Belt; Scheidegger 1986) as well as from the South (Sierra Madre and Pacific Coast of Oaxaca State E to Tehuantepec) and from Yucatan (Scheidegger 1989); the results are given in Table 2.3 and in Fig. 2.4. An examination of these results shows a systematic variation of the joint and stress orientations: In the Center and in the South the direction of one of the bisectrices, presumably the maximum compression (P), is normal to the trend of the coast; as the latter turns, the pressure direction also turns. In effect, the coast is parallel to the margin of the Cocos Plate and the corresponding supposed subduction zone; the compression lies naturally in the direction of the subduction. This interpretation is confirmed by an analysis of fault plane solutions of earthquakes (cf. Scheidegger 1989).

An analogous situation is encountered in Yucatán: One bisectrix, presumably P, follows the coast, oriented normal to it. Naturally, the coast is not a plate margin; the boundary between the North American and Caribbean Plates follows a transcurrent fault whose strike lies at 45° to the maximum compression: thus the latter must be directed EW in the adjacent North American Plate, as was inferred from the joint orientation measurements.

2.2.3.6
Adjacent West Atlantic Islands

General Remarks Adjacent to the North American coast in the East, there are several archipelagoes, such as the Bahamas (Scheidegger 1977). The Caribbean islands are usually assigned to a special plate, apart from Laurasia, and will be treated together with South America.

Bahamas The Bahamas form an archipelago which extends roughly from 21° to 27° N latitude and 73° to 80° W longitude. It consists of 29 inhabited islands, 661 keys, and 2387 pinnacles rising to within 6–7 m below the sea surface, reaching from just E of Florida to just N of Española. The Bahama islands are the topmost parts of a shallow oceanic plattform (the Bahama Bank) which rises from the abyssal plain to about sea-level. A deep test well in Andros Island revealed Lower Cretacteous reef carbonates at a depth of 4456 m, from which it is concluded that the major part of the Bahama Platform was, since the late Jurassic or early Cretaceous, a site of subsidence and carbonate sedimentation (Weyl 1966, 1975). The solid surface in the Bahamas is made up largely of shallow marine deposits, littoral deposits, and wind-blown material (eolian deposits). It presents a mostly flat aspect with some low ridges whose highest point is 68 m above sea level.

Joint orientation measurements have been made on four islands (New Providence, Bimini, Eleuthera, Grand Bahama) in the N of the archipelago. The results are given in Table 2.3 and in Fig. 2.4. It is seen that, whilst the scatter between the various islands is rather large, the joint orientation pattern for the whole archipelago fits together with that common in N America.

Interpretation Considering the data from the various islands, it should be noted that those investigated are not truly "oceanic" but are all adjacent to the nearby continent. It is no surprise, then, that the joint/stress orientations are essentially connected with the plate tectonics around the continent in question.

2.2.4
Laurasian Arctic and Subarctic Regions

2.2.4.1
Introduction

The arctic and subarctic land regions all belong to Laurasian continents or lie on their continental shelves. From east to west, we are concerned here with Spitsbergen, Greenland, Baffinland, the Hudson Bay region and the Great Slave Region.

2.2.4.2
Spitsbergen

General Remarks Commonly, the name "Spitsbergen" has been applied to the archipelago consisting of 5 large and many smaller islands stretching from

76°26' to 80°50'N and from 10°30' to 28°12'E (although this name has lately been restricted to the largest island of the group, which was formerly called "Vest Spitsbergen"). The archipelago lies on an extension of the N European continental shelf.

Our studies were made from the headquarters at Longyearbyen (where daily flights land), from where several day-excursions into the surrounding region could be made as far as roads existed. In addition, a cruise was taken on a regular mail ship which led as far north as Moffen Island (lat. 80°06'N) and back, with stops in the Magdalene Fjorden, at Ny Alesund and at Barentsburg. Around Longyearbyen and during the time ashore, outcrops were inspected and joint orientations measured.

Geology   Stratigraphically, most formations from late Precambrian to recent occur in Spitsbergen. The oldest rocks appear on the W and N: Zircons, granites and gneisses have been dated up to 3200 Ma ago and have been involved in mountain building and metamorphic episodes approximately 1700, 1000 and 600 Ma ago (Hielle, 1993). Next follow (Orvin, 1966) series of Precambrian, Cambrian and Ordovician dolomites, shales and quartzites. All these old rocks have partially been involved in the Caledonian orogeny at the end of the Silurian time 400 Ma ago and form the Caledonian folds and overthrusts of the West. Outside of these old rocks there is a plateau of relatively undisturbed strata on Vest Spitsbergen, lying unconformably on a plattform of pre-Devonian crystalline rocks. In the NW there are Devonian rocks. These are unconformably overlain by Carboniferous sedimentaries with some coal, followed by sandstones, limestones and shales dating successively from all ages, from Permian to Cretaceous (Orvin, 1966). During the Cretaceous, Spitsbergen was uplifted, so that the Upper Cretaceous beds have been eroded and a Tertiary conglomerate, less than 1 m thick, lies unconformably upon the Lower Cretaceous. Further Tertiary sequences consisting of sandstones and siltstones, with some shales, overly it; in its lower parts, important coal seams are found (Hielle, 1993). An extinct volcano and several hot springs (28°C) in the north are Quaternary. The strand flat is well developed to 55 m, and postglacial raised beaches are marked (Orvin 1966).

Morphology   About 58% of Svalbard is covered with ice. The mountains at the W-coast of Spitsbergen are rather wild with sharp ridges and peaks formed of folded and metamorphic pre-Devonian sediments. The E part of Spitsbergen has plateau-formed mountains built up of nearly horizontal younger sediments. In the N part of Spitsbergen there are more rounded mountains built up partly of granites and gneisses. Many large fjords penetrate the west and north coasts of Spitsbergen; important are Isfjorden (with many branches, such as Adventfjorden), Kongsfjorden, and Magdalenefjorden on the W-coast. Some glaciers do not reach the sea, giving rise to ice-free valleys, such as Sassendalen, Adventdalen, Colesdalen and Reindalen (Orvin, 1966).

Locations visited[1]

Adventdalen. Adventfjorden is a side fjord of the large Isfjorden; it is continued by the Adventdalen, which is an ice-free valley containing the capital Lonyearbyen, with many side valleys, all ice-free. The only connected road system of Spitsbergen exists here, so that the side valleys mentioned above could be visited by car. The bedrock consists on the lower slopes of Lower Cretaceous fluvial and deltaic deposits with sandy beds. Over 200 m of Cretaceous beds are exposed near Longyearbyen. Joint orientation measurements were made at 6 locations in this area mainly in Tertiary shales, but also in Cretaceous shists and limestones.

Grönfjorden. Grönfjorden is a further side fjord of Isfjorden, on its S-side not far from its mouth and to the W of Adventsdalen. On the E-shore of Grönfjorden lies Barentsburg, a Russian mining town. At the E entrance of Grönfjorden, Jurassic and Cretaceous rocks with many fossils of dinosaurs are found: The town of Barentsburg lies in it. The coals, however are found in the overlying Tertiary (Hielle, 1993, p. 54–55). Joints were measured in this fiord at two outcrops, in Cretaceous shales and limestone layers.

Kongsfjorden. A landing was made at Ny Alesund. The coast at this location consists entirely of Carboniferous to Permian strata. The landscape is plateau-like open tundra, on former beach ridges. Joints were measured at 2 outcrops near Ny Alesund in Carboniferous-Permian limestones.

Magdalenefjorden. Another landing was made in the Magdalenefjord. The rocks belong to the crystalline basement (Proterozoic granite and mica schist with carbonate and quartzite beds), forming rugged peaks. Several large glaciers calve into the fjord in the vicinity. Two outcrops were found at which joint orientations could be measured, both in Proterozoic granite.

Discussion

An inspection of Table 2.4 shows that all joint directions correlate pretty well with each other in the various areas. There is no effect of the lithology on the joint orientations; however, joint strikes are generally parallel to the directions in which the various fiords are trending: The orientation data of the 249 joints mentioned above were compared with the trends of the fiords. The latter were segmented and evaluated according to the usual statistical method; the results are listed in Table 2.4 as well. There were basically two conjugate fiord sets whose trends (statistical maxima at N77°E and N161°E) are close to the two conjugate strikes of the joints: the E-W striking trends are only 9° apart, which, within the error limits, can be regarded as a reasonable fit.

This has a bearing upon the theories of fiord genesis. "Glacial Overdeepening" has generally been held responsible for the genesis of fiords (Holtedahl 1967). A completely opposite view has been put forward by Gregory (1913) who maintained that fiords are of tectonic origin and that the glacial influence was negligible during their genesis. Holmes (1944, p. 224) combined the two views

---

[1]Map in Hantke and Scheidegger, 1999

Table 2.4. Laurentian Arctic: joints, fiords, strike/trend directions

Location	No.	Max 1	Max 2	Angle	Bisectrices	
Spitsbergen						
Adventdalen	144	166±07	81±08	85	123	33
Barentsburg	21	178±12	80±15	81	39	130
Ny Alesund	42	13±09	84±13	71	49	138
Magdalenefjor.	42	32±12	106±15	74	69	159
All Joints	249	179±01	86±03	86	42	132
Fiords	66	161±03	77±12	84	119	28
Greenland						
Joints	102	140±12	47±05	88	3	93
Fiords	91	136±10	49±02	87	3	93
Baffin Land						
Joints	133	132±00	51±07	81	91	2
Fiords	46	140±03				
Churchill Man	24	96±11	7±15	90	52	141
Great Slave Region						
Yellowknife	152	152±12	57±10	84	15	104
Ft.McMurray	608	139±01	50±00	89	95	5

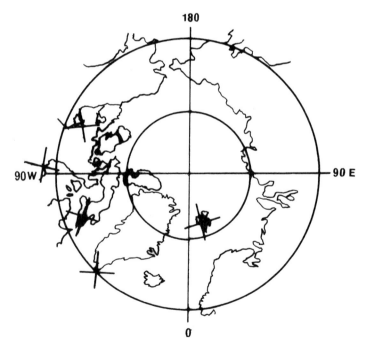

Fig. 2.5. Joint strikes in the investigated Laurasian Polar Regions

and admitted that, in addition to glacial action, an "appropriate structure" of the respective region may have been involved. He also noted that in plan, fiords everywhere have a rectilinear pattern, which he claims to be "clearly determined" by belts of structural weakness, such as synclines of weak sediments or schists enclosed in crystalline rocks. In this instance, the coincidence of fiord trends and joint strikes in Spitsbergen points towards an identical, viz. a neotectonic origin of the fiords (Hantke and Scheidegger 1999).

2.2.4.3
Greenland

Turning now to arctic regions which are near North America we first consider Greenland, where the south coast near the town of Narssarssuaq (latitude 61°10′N and longitude 45°30′W) has been investigated. Outings were made to Brattahlid, Igaliko and the inland ice cap to the NE of Narssarssuaq.

Geology Most of the knowledge of Greenland geology derives from a study of its coasts and extrapolations therefrom. Thus, the greater part of Greenland seems to be an extension of the Precambrian Shield of North America along its northern and eastern margins. This shield area is bounded by Paleozoic fold belts. The Precambrian basement consists of three major units: on either side of the central Archean block, two younger mobile belts of similar age have been recognized. The central (Archean) block extends on the west coast (which mainly concerns us here) from 61°45′N to 66°30′N. The dominant rocks are granulite gneisses of grano-dioritic to quartz-dioritic composition with several successions of metasedimentary and metavolcanic rocks. In our area, the gneisses and supracrustals are cut off by extensive areas of granite and granodiorite which was emplaced during several different periods beginning about 2600 Ma ago. The active period ended with the intrusion of post-kinematic alkaline rocks (norite, diorite, granite), ending about 1780 Ma ago (Bondam 1975, Langel and Thorning 1982).

Joint orientation measurements A total of 101 joint orientations were measured at 5 locations in the general vicinity of Narssarssuaq. The outcropping rocks were mainly Precambrian gneisses and schists. The measurements were then evaluated according to the ususal statistical method; the results are shown in Table 2.4 and in Fig. 2.5. Interestingly enough, similar joint orientations were found in the S of Greenland (Scheidegger 1993a) as on the European Atlantic Seaboard: one of the bisectrices, presumably the maximum pressure direction, was found at N 003 E (cf. Table 2.4); i.e. essentially NS. Since, however, Europe and Greenland are on different sides of the plate boundary represented by the mid-Atlantic ridge, a direct connection between the measurements mentioned can probably not be construed (Scheidegger 1993a).

Morphotectonic conclusions Quaternary glaciation lead to the formation of the Greenland Ice Cap 2–4 Ma ago. Advances to the coast took place during the ice ages; since the last ice age there has been an isostatic uplift of some 150 m. The local morphology has been greatly afffected by glacial action, which resulted in the formation of large fiords which show a grid-like pattern. In order to confirm this visual impression, we have digitized the median lines of the fiords in steps of 2.5 km and treated the orientation data with the usual statistical method. The results of this procedure is also shown on Table 2.4. It is observable that the values of the fiord directions are very close (within 4°) to those of the joint directions: The fiords as well as the joints strike NW-SE and NE-SW. Like in Spitsbergen, this indicates a tectonic (and not by "overdeepening"-caused erosion) origin of the fiords.

2.2.4.4
Baffin Land

Geography
Baffin Island is the south-easternmost island of the Canadian Arctic Archipelago. It comprises an area of 474,000 km^2, has high ($>$ 2000 m) glaciated mountains in the North and a plateau (with tundra vegetation) in the South. For the morphotectonic studies, the area around its capital, Iqaluit on Frobisher Bay, was visited; it lies at a latitude of 63°45'N and a longitude of 68°31'W, about 2000 km north of Montreal. Studies were made around Frobisher Bay and the hills around.

Geology   Frobisher Bay lies in the Churchill Province of the Canadian Shield. The Churchill Province consists of Archean and early Proterozoic sedimentary and plutonic rocks, mainly metamorphosed and converted to granitic gneisses during the Hudsonian orogeny (1.65–1.85 Ga ago) which are often biotitic and interbanded with felsic to mafic gneisses and schists containing variously amphibole, hypersthene, epidote and chlorite. These massive rocks vary from alaskite to ultrabasic rock and include bodies of granulite and charnockite on southern Baffin Island (Christie 1975).

Tectonics   The system of channels that now separates the Arctic islands is evidently a drowned valley system modified by Pleistocene glaciation. The larger channels, many straight and arcuate-walled, are controlled by graben or rift-valley structures: such a structure also separates West Greenland from Baffin Island and formed Baffin Bay. It developed in late Cretaceous/early Tertiary time and is presumed to have been an offshoot of the Mid-Atlantic Ridge. In Baffin Bay, rifting has occurred when the climate was subtropical to warm-temperate (Andrews et al. 1972) resulting in accelerated fluvial erosion along the steep, seaward facing flank of the rift. The variation in land levels from island to island suggests that the major, now drowned, drainage system is also fault-controlled (Christie 1975).

Orientation measurements   Joint orientations were measured at 7 locations around Iqaluit, all in the banded granitic gneisses of the Churchill Province of the Canadian Shield. The data and the values were evaluated according to the usual statistical method; the results are summarized in Table 2.4. Comparing this with the values from W Greenland, one sees at once that the results for

Greenland and Baffin Island are practically identical. This would also correspond exactly to the view that there is an ongoing rift between Greenland and Baffin Island across Baffin Bay, i.e. a stress relief in the direction of E-W; this corresponds to a T direction also trending E-W (Scheidegger 1998a).

Furthermore, one joint set strikes exactly along the trend of the local valleys (Jordan River, Sylvia Grinnell River, Niaqunguk River, Armshow River etc. and last but not least Frobisher Bay itself). There are therefore many indications that these valleys have been tectonically predesigned.

2.2.4.5
Hudson's Bay

Morphotectonic studies were made in the region of Churchill (58°48'N latitude, 94°10'W longitude) on Hudson's Bay. Outcrops were investigated around the coastal ("Launch") road, on Cape Merry at the mouth of the Churchill River and near the Churchill Northern Studies Centre.

Geologically, the region around Churchill lies on a Precambrian Basement. However, in the low parts, the Precambrian is unconformably overlain by Ordovician limestones. Morphologically, there are many long ridges present, called "eskers", generally followed by the roads in the area; however, these are not eskers in the geomorphological sense, but rather ancient beach ridges that have been raised by isostatic uplift in the wake of the disappearance of the ice of the last glacial age. On the other hand, kames are frequent (e.g. Twin Lakes).

A total of 24 joint orientations were measured in the area at the outcrops indicated. The measurements were then evaluated according to the usual statistical method. In summary the results are given in Table 2.4. It is interesting to note that this correlates exactly to the directions found generally in Central Canada (see Sect. 2.2.3).

2.2.4.6
Great Slave Region

General Remarks   The region around Yellowknife, capital of the Canadian Northwest Territories, situated at 62°27'N latitude and 114°22'W longitude, was studied with regard to morphotectonics. Fig. 2.5 shows its general location in Canada. It abuts in the South on a bay of Great Slave Lake, to the N it borders a small lake, Frame Lake. A marked trail leads around it.

Geology/Morphology   Yellowknife lies in the Canadian Shield, in the province named after it (Wilson et al., 1956; also called Slave Province by Stockwell, 1975), which is one of the oldest of the entire Shield (> 2,560 Ma; Stockwell 1975). Thus it belongs to the Archean part of the Canadian Shield which was a period of much volcanic activity and of the deposition of immense thicknesses of sediments well preserved in the Slave Province where they were reworked by the Kenoran orogeny. Accordingly, flow folds and dykes are common in the corresponding metasediments and metavolcanics.

The area was much affected by the last ice age, thus there are, morphologically, many glacially sculptured hummocks consisting of Archean rocks and
corresponding hollows with ponds. The vegetation consists of boreal forest
and mosses in the wetlands.

Joint Orientation Measurements   Joint orientation measurements were made at
7 locations in the city itself and along a trail leading into the bush east of it.
All were in Shield rocks. The values of the joint orientations were evaluated
in the usual manner; the results are shown in Table 2.4. Only the statistics
for all joints combined deserve any real confidence. It is difficult to make
regional comparisons of these values with others; Babcock (1975) had measured fracture phenomena (not only joints, but also many features from air
photographs) in the Ft.McMurray region which were subject to a statistical
analysis (Scheidegger 1983b). The results are also listed in Table 2.4. Inasmuch
as the features measured by Babcock differ from those investigated around
Yellowknife, no proper comparison is actually possible; nevertheless, there is
a correspondence: the joint strikes agree within the error limits.

2.3
Gondwanaland

2.3.1
Africa

2.3.1.1
Morphology/Geology

Africa borders against Europe in the north; it is separated from the latter by the
Mediterranean Sea. However, the plate boundary between Africa and Eurasia is
not clearly defined; it has been identified with the "Periadriatic" (or Insubric)
lineament to the north of the Mediterranean (see the Section on Europe) or
along a line connecting some of the major islands (the Balearic islands, Malta,
the South-Aegean Arc). The Atlas Mountains are evidently also part of the
Alpine-Himalayan chain and should actually be counted to Europe. However,
we shall adhere here to the conventional geographic separation which considers
the Mediterranean islands as part of Europe and "Africa" to be bounded by its
coastline. One usually takes the Suez Canal as Africa's boundary towards Asia;
but this is also rather artificial inasmuch as the Arabian Peninsula (including
Sinai) is geologically part of Africa. In this book, the Sinai Peninsula will be
treated together with Africa. In the east and west, the coast lines towards the
Indian and Atlantic Oceans are taken as the boundaries of Africa.

The nucleus of Africa is a Precambrian shield, consisting of intrusive crystalline and ancient metamorphic rocks. Upon the shield as a basement, many
sedimentary sequences rest whose materials may be derived from beyond
the present Africa. Structures found in these continue into South America,
which gave rise to the hypothesis of the existence of an original supercontinent
of Gondwanaland (cf. Sect. 1.6.2). Extensive volcanism was (and is) widely

present: most of the rock systems show traces of associated volcanics, some on a very large scale. Notable are the subrecent floodbasalts of Ethiopia and the recent volcanic activity in the East African Rift Valley (King 1966).

2.3.1.2
Egypt

Egypt forms the NE-corner of Africa, stretching from the Mediterranean coast (ca. 31°N) to the frontier with the Sudan (22°N) in the N-S direction and from a line at 25°E in the Libyan Desert to about 37°E on the coast of the Red Sea in the W-E direction.

Egypt is almost wholly a part of the Gondwana foreland, the entire country being underlain by a crystalline basement complex. Upon these old Precambrian rocks have been laid down a succession of later rocks, younger to the north. All the geological eras are represented, at least in part, with sandstone of Mesozoic age ("Nubian Sandstone") and (less so) Eocene limestone ("Thebes

Table 2.5. Joint and river directions in Africa, Strike/Trend directions

Location	No.	Max. 1	Max. 2	Angle	Bisectrices	
Egypt						
Cairo-Sinai	11	107±08	179±21	72	143	53
Kharga	6	103±12	0±10	77	142	51
Thebes	21	161±21	90±14	70	126	36
Abu Simbel	6	108±06	176±43	68	142	52
All Egypt	45	111±07	172±17	71	136	46
Ethiopia						
Flood prov.	41	47±23	154±16	74	10	100
Rift valley	42	9±13	89±13	80	49	139
All Ethiopia	83	2±13	90±16	89	46	136
S-Sahara	382	139±05	50±06	90	95	5
Nigeria						
Nigeria av.N	4*	164±06	83±12	81	123	33
Nigeria av.S						
Joints	2*	131±13	36±17	85	173	83
Riv. Niger/Benue	2	145	65	80	15	105
All Nigeria jo.	6*	165±20	83±20	82	124	33
South Africa						
(1)West	2*	150±05	58±04	88	14	104
(2)East	8*	179±15	93±13	86	136	46
Mine	139	3±09	110±05	76	147	57
Ave. all groups	10*	169±09	83±18	86	126	36

*areas containing some scores of measurements each

Formation") covering the greatest surface area. Pleistocene sediments are of lesser importance (Mountjoy 1966; Said 1962).

Joint orientations (Scheidegger 1995) were measured (a) in the Cairo Region, characterized mainly by a middle Eocene formation, and on the Sinai Peninsula consisting of heavily faulted and folded Recent to Precambrian rocks, (b) in Middle Egypt in Eocene rocks, chiefly limestones, (c) in the Kharga Oasis (a depression lying about 200 km West of the Nile) which is bounded by steep escarpments of a huge limestone plateau whose floor consists of Nubian Sandstones (see above); and finally (d) in Upper Egypt (Abu Simbel) which is characterized by vast extensions of the Thebes Formation (see above) and Nubian Sandstones.

The data from the several regions mentioned were evaluated according to the usual method (Sect. 1.3.3); the results are listed in Table 2.5 for the individual locations as well as for all joints taken together. It is interesting to note that the joint strikes resulting from these measurements are quite consistently oriented NS and EW. The bisectrices of the joint maxima have orientations of NW-SE and NE-SW. The NS joint strike direction correlates with the direction of the Nile Valley between the Mediterranean coast and Middle Egypt; the latter seems therefore to be tectonically predesigned in the same manner as the joints. Above (i.e. south of) Middle Egypt, the Nile changes to a NW-SE direction, which is also parallel to the trend of the Gulf of Suez–Red Sea. This agrees with the direction of one of the bisectrices of the joint maxima. If the latter is interpreted as the greatest compression, then the greatest stress relief is oriented NE-SW, which is roughly normal to the trend of the Gulf of Suez–Red Sea. This would be in conformity with the idea that the latter is a *tensional* "graben".

2.3.1.3
Ethiopia

Ethiopia encompasses mainly the highlands around its capital Addis Ababa (9°00′N, 38°44′E), with an area of 1.06 million km^2 compactly situated within ca. 15° of latitude and 15° of longitude. It is a tableland consisting of Precambrian basement rocks covered by sedimentaries and large areas of great thicknesses of volcanics: In the North, the volcanics occur as a province of flood basalts; in the South they are the result of recent volcanic eruptions in the Ethiopian branch of the East African rift system. The volcanic activity was/is associated with extensive faulting connecting with rifting. In the North voluminous outpourings of lava have built stacks of superimposed basalt flows, resulting in the formation of a subdued topography, which, due to neotectonic uplifting, results in the aspect of a "volcanic plateau" ("Flood Basalt Province"). The basalts are transitional between alkaline and tholeiitic (Mohr 1983). The rift zone is characterized by a tectonic moving-apart and contains many volcanoes; the volcanism has started some Ma ago, culminating about 0.2 Ma ago, but is continuing to the present day (Mohr et al. 1980). The rift volcanoes are typical shield volcanoes, generally 15–20 km in diameter and 200–800 m high; most have a summit caldera caused by the collapse of the magma chamber below. The volcanoes are set between offset segments of fault belts paralleling the Rift, which have been proven by geodetic measurements to be tectonically active to the present day. Their lavas are mainly basaltic, sometimes rich in carbonates (Ollier 1981, p. 112).

Studies of joint orientations (Scheidegger 1982b) have been made in lavas on the flood basalt province around and north of Addis Abeba as well as in the Ethiopian rift. If the genesis of the joints had been connected with the emplacement of the lava, their orientation should be random. This is, however, not the case: The orientation of the joints in the rift-outcrops is practically identical to those determined for Ethiopia as a whole and are also in the same quadrant as those for the flood basalt province; the numerical results are shown in Table 2.5. If the joints indicate shear lines in a (neo)tectonic stress field, then the bisectrix-directions would be parallel and normal to the strike of the rift (striking N 40°E just S of Addis Ababa) and of the faults would parallel the latter. This would fit the hypothesis of a neotectonic stress relief (i.e. tension) normal to the rift, as has also been established by fault plane solutions of earthquakes (Kebede 1989). Thus, it appears that the joints in the volcanic outcrops in the Flood Plateau and in the Rift are not cooling cracks of lava, but have been mechanically predesigned by the (neo)tectonic stress field in the area.

2.3.1.4
Southern Sahara

Joint measurements have been reported (Erlach and Scheidegger 1983) from the border area of Algeria, Niger and Libya (ca. in a square 22°–28°N and 4°–10°E). The region concerned lies geologically at the boundary between the Precambrian of the African Shield in the S and the oil producing Paleozoic cover-strata further North which come to the surface here. Four groups of outcrops were investigated which showed sandstones, limestones, quartzites and cross-bedded Silurian siltstones (Legrand 1981).

The evaluation of the data yielded quite similar results for all four groups of outcrops: The strike directions of the joint sets are essentially NE-SW and NW-SE. The strike of the second of these sets agrees with the trend of the nearby Tassili-n-Asdjer mountains which points to a morphotectonic relationship. If the joints are interpreted as shearing phenomena of a neotectonic stress field, the principal stress directions (bisectrices) would be roughly EW and NS. If the second of these is taken as a maximum pressure direction, this could result from the general northward wander of the African plate and the ensuing collision with Europe. The numerical results of the evaluations are given in Table 2.5.

2.3.1.5
Nigeria

In Nigeria, several separate morphotectonic studies have been made (Scheidegger and Ajakaiye 1985, 1990; Ajakaiye and Scheidegger 1989; cf. also a summary in Scheidegger, 1995).

Geologically, Nigeria is characterized by Precambrian basement complexes consisting of migmatitic schists and old metamorphics; in some places they are intruded by "older granites" (600±150 Ma) that originated in the Pan-African orogeny. The basement blocks are separated by a Y-shaped boundary formed by the Niger and Benue valleys; the latter are now filled with Cretaceous to Tertiary sediments. More recently, an uplift occurred around Jos in consequence of Jurassic intrusions of "younger granites" which appear as ring dykes representing old calderas of volcanoes (Scheidegger and Ajakaiye 1985).

In our morphotectonic studies it turned out that Nigeria is divided into two regions: north of the Benue trough and south of and including the Benue trough, i.e. N of 9°N and S of 9°N. In each of these zones, the joint orientation measurements at the respective outcrops gave essentially consistent results with a substantial change occurring at the 9°N-boundary (cf. Ajakaiye et al., 1988; Scheidegger and Ajakaiye 1990; Scheidegger 1995). Thus Table 2.5 shows averages for the areas north and south of 9°N. In view of the difference between North and South, it is evidently not really meaningful to calculate an average result for Nigeria as a whole.

In summary, one can state the following: The joint and stress directions are very consistent amongst each other north of 9°N and south of 9°N. To the north of 9°N, the prevalent joint sets strike more or less NS and EW, the bisectrices are NW-SE and NE-SW; these are the same directions as in the African Rift in Ethiopia which would fit a plate tectonic model in which there is a moving-apart of crustal blocks at the Rift (tension at the plate margin) leading to a compression in the same direction within the African plate. To the south of the 9°N as far as the Benue Trough and the Cameroon Mountains, the joints/stresses are quite different from those to the north (in effect they are turned by about 50° with regard to the latter): the joint strikes are parallel to the trend of the Benue trough (N65°E) which is known to represent a major shear zone (Pelusium Megashear) that finds its continuation on the other side of the Atlantic in Brazil. This trend is approximately conjugate (angle 80°) to the trend of the Niger valley (N145°E) above the confluence with the Benue. The trend of the Niger agrees with the second joint strike maximum in Southern Nigeria. Evidently, the main river trends in southern Nigeria correlate more or less with the joint strikes and thus may be supposed to be related in their morphotectonics.

2.3.1.6
South Africa

Morphotectonic studies were made in the triangle between Johannesburg, the Transkei and Durban. Regarding the generalized geology and geomorphology of this area (Kent 1980; Truswell 1977) one may note that the north-eastern regions (north of Potchefstroom and east of Johannesburg) are part of the Precambrian shield, consisting of the "Bushveld complex" and "Transvaal sequence". The Bushveld rocks are norite, granophyrite and granite, 1670–2100 Ma old (Kent 1980). The landscapes vary from flat to slightly undulating and are situated at an elevation of approximately 1350 m above sea level.

The Bushveld complex is surrounded by the rocks of the Transvaal sequence into which it intruded. The Transvaal sequence consists of meta-sediments (age 2224±21 Ma [Kent 1980]) and diabase-intrusions. The geomorphology of this area is controlled by the parallel to subparallel northward dipping strata leading to a series of homoclinal ridges with their intermediate valleys. The altitude of this area increases from the floor of the Bushveld basin (1350 m) to 1600 m above sea level. The region to the west of Johannesburg (West Rand) is a high field at an elevation of approximately 1500 m consisting of Archean granite (2320±65 Ma old, Kent 1980) and younger intrusions of uncertain age. The landscape is a flat to slightly undulating plain with isolated tors. The region around a strange dome-structure (Vredefort Dome near Parys [26°53′S, 27°28′E]) consists of a series of ridges in a semi-circular form surrounding the northern sector of the granite. Nicolaysen and Ferguson (1981) claim (in view of the lithology) that the Vredefort basement rocks were uplifted by at least 30 km. The uplift mechanism is still a matter of controversy: a doming process (Du Toit 1939) or crustal rebound after an impact event (Dietz 1961) has been advocated as its cause. The area is at a general altitude of between 1290 m and 1500 m. The rest of the investigated regions are all on different members of the so-called Karoo Sequence which was (Truswell 1977) gradually deposited from the late Carboniferous to the early Jurassic in a shallow, dry (Karoo) basin which comprised a large part of southern Africa.

Joint orientations were measured in key areas from Johannesburg-Pretoria in the North to the West Transkei (near East London) and the Barkly Pass in the SW of the country. In addition, Prof. J.T. Harmse contributed data from 139 joints in a mine in the Rand 2187 m below ground (personal communication). These key areas fall into two regional groups, in each of which the joint orientations turned out to be consistent: (1) the "West", encompassing the outcrops in the Western Transkei on the coast around Mazeppa Bay [32°21′S, 28°47′E] and those around the Barkly Pass [near Elliott, 31°58′S, 28°42′E]; and (2) the "East" encompassing the rest of the country from the Rand to the Eastern Transkei. The data were evaluated according to the usual statistical method (Sect. 1.3.3); for each key area two "preferred" near-vertical joint orientations were determined: the latter were then assigned to the two basic regional groups. The bisectrices were considered as the hypothetical principal directions of the stress field causing them. A detailed description of the results has been given by Scheidegger (1995); Table 2.5 gives a summary of the results for the two regional groups; the situation is also depicted in Fig. 2.6.

In summary, it may be stated that the orientations (strikes) of the joints in the "East" of South Africa are oriented mainly N-S and E-W. There is a consistency of joint orientations from the Bushveld to the Eastern Transkei and Natal. It is also worthy to note that the joints measured below ground in the mine have essentially the same orientation as the surface joints in the area. Things are different in the "West" from the Western Transkei to Barkly Pass, so that one might suspect a neotectonic boundary in the western part of the Transkei. The observed consistency of the joint orientations over the hundreds of km from Pretoria to Durban conforms to the observations made elsewhere in the world, where it has generally been found that joint orientations do not change

Fig. 2.6. Joint strikes in various regions of Africa

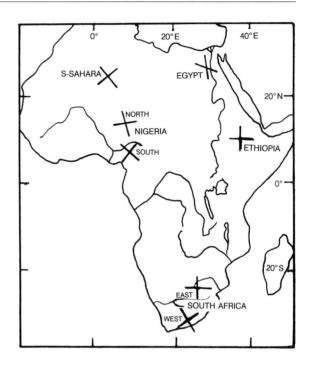

much over large intraplate regions. In an evaluation for all of South Africa, the anomalies of the "West" simply disappear.

2.3.1.7
Summary for Africa

Table 2.5 and Fig. 2.6 summarize the joint/stress situation in Africa: The stress (bisectrix) directions calculated from joint orientations at the eastern rift zones (Red Sea, Central Ethiopian Rift) show one bisectrix of the joint sets, presumably a stress relief (maximum tension) at right angles to those rifts, as one would expect from plate tectonic models. For the rest, the joint/stress orientations are continent-wide very uniform (except for the local anomalies in "Nigeria-South" and "West South Africa"): basically a NW-SE principal stress in the south turning slowly toward a more E-W direction in Libya, representing an anticlockwise rotation. This, again, would be the expected intra-plate tectonic stresses inasmuch as the African plate is generally moving northward towards Europe, but being dragged eastward in the south by the opening of the mid-Atlantic ridge (cf. Sect. 2.4). The two anomalous regions may have been influenced by their location close to the coast. Thus, it turns out once more that the stresses deduced from joint orientations fit the patterns expected from plate tectonic models.

2.3.2
Peninsular India

2.3.2.1
Physiography/Geology

Peninsular India is that part of the Indian subcontinent which lies south of the mountainous regions of the Himalaya. It encompasses the triangle from the great Indo-Gangetic plain of the Punjab and Bengal extending from the valley of the Indus in Pakistan to that of the Brahmaputra in Assam to the south as far as the island of Sri Lanka. It consists mainly of an old (Precambrian) basement, covered in part by volcanic Deccan lavas (emplaced at the end of the Cretaceous) which cover an area of about 500,000 km^2 mainly the NW-central part from 12° to 21°N latitude.

Morphological, structural and seismic information was collected from widely separated areas of Peninsular India, reaching from Delhi and Bombay to the Krishna River valley N of Bangalore. The localities covered were situated in country rock ranging from non-volcanic Precambrian gneisses to Deccan volcanics. The morphological studies involved orientation structures of river courses, the structural studies joint orientations and in-situ stress measurements, and the seismic studies fault plane solutions of earthquakes.

2.3.2.2
Non-Volcanic Areas

Geology The general geology of the Precambrian basement of Peninsular India involves four systems, called (from oldest to newest): Archean (gneiss and schist), Cuddapah (former mountains), Delhi (quartzites, arcose, micaschists), and Vindhyan (sandstones). No absolute ages are available for these systems, but they are all greater than 1 Ma. Overlying the Precambrian are, in places, as noted, Deccan lavas and younger sediments, notably recent alluvium in the Gangetic Plain (cf. Wadia, 1975).

Orientation measurements *Joint orientation measurements* were made in four regions in non-Deccan areas: Delhi, Hyderabad, Achampet and Sri Sailam. In *Delhi,* three outcrops were visited on Nehru Ridge, all showing Delhi Quartzite. In the more *southern parts in Precambrian regions* (Scheidegger and Padale 1982), three outcrops were located near *Hyderabad* in Archean gneisses, further South another three outcrops near *Achampet* again in Archean gneiss, and finally near *Sri Sailam,* in Proterozoic sandstones from the Cuddapah System. The results of the usual statistical evaluations are shown in Table 2.6. It is observable that all results from Precambrian Peninsular India fit reasonably well together with each other, so that makes sense to calculate a mean.

In addition, *in situ stress* measurements have been reported from the Kolar gold fields near Bangalore (Gowd et al. 1981). The prevailing direction of the maximum neotectonic compression was found to be between N130°E to N145EW (say 137E±7E).

Table 2.6. Evaluations for Peninsular India; Strike/trend directions of joints/rivers and bisectrices

Location	# outcr.	joints	Max 1	Max 2	Angle	Bisectrices	
Non-Deccan joints							
Delhi	3	63	16±20	112±11	86	155	64
Hyderabad	3	65	1±14	94±14	87	138	47
Achampet	3	71	74±08	162±34	87	118	28
Sri Sailam	3	66	118±09	17±07	80	158	67
All non-Deccan							
Joints	12	4reg	100±13	4±11	85	142	52
In-situ P						137	
Deccan joints							
Pune region	9	67	94±14	3±14	90	130	49
Aurangabad	3	46	76±20	174±15	83	135	25
Ajanta Caves	2	43	97±18	4±15	87	140	50
Ahmednagar	2	211	107±17	11±30	84	149	59
Karla Cave	3	64	81±19	160±19	80	120	31
Borgaon	3	66	106±26	19±19	87	62	152
Koyna region							
Joints	10	235	84±09	175±11	80	40	130
Rivers nr Koyna			86±03	165±08	79	126	36
Earthquakes T		11					45
All Deccan joints	32	7reg	90±02	00±04	90	135	45
Pen. Ind. ALL joi.	44	11reg	92±02	1±04	88	137	47

2.3.2.3
Volcanic Areas: Deccan Traps

General remarks   The Deccan lavas are a great volcanic formation that had been emplaced towards the close of the Cretaceous era when a large part of the Peninsula was affected by large volcanic eruptions. Highly liquid basaltic lava welled up intermittently from linear fissures, spreading out until a thickness of some thousands of meters of horizontally bedded sheets had been emplaced, obliterating all the previously existing topography and converting it into an immense volcanic plateau. The latter rose to about 600 m mean elevation above sea level and was enclosed on all sides by steep-sided and terraced hill ranges such as the Western Ghats with a mean elevation of some 900 m, running at a distance of about 50 km parallel to the Arabian Sea coast. Whilst the stepped and dissected morphology of the Deccan Traps has mostly been considered as due to erosion, it is evident that such features as the Western Ghats have a geotectonic origin by scarp faulting (Wadia, 1975). Thus, the Deccan Traps are a prime example of the presence of mechanical designs in volcanic landscapes characterized by *basaltic* volcanism.

Fig. 2.7. Preferred joint strikes in Peninsular India

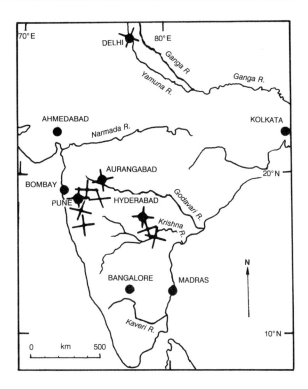

Orientation Measurements To substantiate these views, a geodynamic study of pertinent features (joints, valley directions, earthquake faultplanes) was made in central India (Scheidegger and Padale 1982) which yielded the result that the orientation structure of the geomorphic features is nonrandom and therefore of non-exogenic origin.

Thus, *joint orientation measurements* were made at nine outcrops near *Pune* (Poona), in the city itself and in its neigborhood, near Khadakwasla Lake and near Wagholi. The outcrops were found mainly in lava bands. Towards the NE of Pune, three outcrops were visited near *Aurangabad*, two at a waterfall near *Ahmednagar* and two at the *Ajanta Caves*, all in typical trap-basalt, sometimes intercalated with welded tuffs. The caves were situated in rather interesting "erosion cirques". On the seaside of the Western Ghats, the region around *Karla Cave* was investigated (three outcrops); the rock was again typical trap-basalt. South of Pune, a group of three outcrops were studied near *Borgaon*, on road cuts in typical trap-basalt. The largest effort was concentrated on the *Koyna area*. Here, damage had been done by an earthquake to a water reservoir (lake) impounded for the purpose of hydroelectric power generation. The earthquake had been thought to have been triggered by the engineering activity in the region, but it was suspected to have been tectonically codesigned. Thus, the region which extends from the lake across the Ghats (Kumbarli Pass) to the coastal plane near Chiplun was carefully investigated. Not only were joints measured at ten outcrops, but the *valley trends* were also noted. The joint orientations and valley trends were subject to statistical analyses; the results are shown in Table 2.6. In addition, the earthquake fault plane solutions of the Koyna earthquake of 10 December 1967 and subsequent earthquakes in the area were noted and statistically processed: only the trend of the tension(T)-axis was reasonably defined ($T = N45° \pm 31°E$); it is also listed in Table 2.6. All the features indicated above were evidently statistically correlated: the joints, the stream valley and the Koyna earthquakes seem to have a common neotectonic origin.

2.3.2.4
Summary for Peninsular India

It remains to make a comparison between the joint orientations inside and outside the Deccan traps. As shown above, joint orientations were measured in 4 regions (12 outcrops) outside and in 7 regions (32 outcrops) inside the Deccan traps. The "Deccan" and "Non-Deccan" maxima of joint orientations have then been calculated; the corresponding numerical results are given in Table 2.6. This may even be correlated with the river segment orientations near Koyna and the in-situ measurements in the Kolar gold fields which fit exactly with the results from the Koyna earthquake fault plane solution. Thus, it is quite evident that there is no difference between Deccan- and Non-Deccan regions. This means, of course, that the patterns in question are caused by tectonic and not by intrinsic lava-genetic or exogenic processes: the designs are mechanical and not epigenetic.

2.3.3
Australasia

2.3.3.1
General Morphology/Geology

"Australasia" is commonly defined as encompassing continental Australia (with Tasmania), New Guinea, the islands of Melanesia and New Zealand. All these regions had a common geological evolution. The tectonic development of Australasia began around an early Precambrian (6000 Ma old) shield in Western Australia to which Eastern Australia was accreted in the Paleozoic. Subsequently newly developing orogenic zones migrated to the NE until the island complexes of New Guinea, the Solomons, Vanuatu (formerly called New Hebrides), Fiji, New Caledonia and New Zealand were formed in the Mesozoic and Cenozoic (Rickard 1975). Morphotectonic studies were made in Australia, New Zealand and Fiji (Scheidegger 1995).

2.3.3.2
Australia

General Remarks Australia is an island continent covering 7,667,080 km^2 stretching from 10°S to 44°S latitude and from 113°E to 153°E longitude. As noted above, the continent of Australia consists of a nucleus or shield of ancient Precambrian rocks mainly in the West, the far South and in the Northern Territory, but Precambrian rocks occur in all states except Victoria. The eastern states were built up by the "Tasman" orogeny in the Paleozoic evident in the mountains of New South Wales and Victoria. Between the shield and the remnants of the Tasman orogeny a depression (Central Depression) was formed in the Mesozoic. The "ranges" therein (MacDonnell, Olgas) are erosion remnants of mildly deformed regions of the depression (D'Addario 1975). Joint orientation measurements have been made in the South-East (Locations at Canberra, NSW-Coast, Snowy Mountains, Victorian Alps, Gambier, Victoria Coast and Camperdown), the Central Depression (Alice Springs, MacDonnell Ranges, Olgas) and in the Top End (Darwin) (Scheidegger 1980).

Southern Australia Four regions were investigated in Southern Australia: (1) In the *Canberra Region* around Canberra, the capital of Australia. The latter is situated at 45°21′S latitude and 149°10′E longitude in a broad depression of a tableland. The rocks are Ordovician sandstones which have been strongly folded by a Paleozoic orogeny. Joints were measured on the two most prominent hills in the region: Black Mountain and Mt. Ainsley. Furthermore, another outcrop was studied at the banks (consisting of Ordovician sandstones) of the Murrumbidgee River 20 km South of the city. (2) At the *New South Wales Coast* three outcrops were investigated near Bateman's Bay (54°50′S, 150°10′E) on cliffs consisting of Ordovician shales and limestones. (3) In the *Snowy Mountains*, three outcrops were studied around Mount Kosciusko (36°17′S, 148°30′E), all in so-called "Kosciusko Granite", an intrusion-batholith in Silurian continental shelf deposits. (4) In *South-Central Australia,* four outcrops were inspected and joint orientations were measured (i) in the Victorian Alps (mainly Devonian terrestrial sandstone with only very simple structures); (ii) at Mt. Gambier and Camperdown (very recent volcanics, 5000 – 7000 years old, waterlaid tuffs and cinders) and (iii) along the Victoria Coast (mainly in Tertiary limestones). The joints of "Southern Australia" were *evaluated* according to the standard statistical procedure; the results are shown in Table 2.7. An inspection of the latter shows that there is a great consistency regarding joint strikes throughout the area, except at Canberra. It is not uncommon in the world that a small anomalous areas do not fit into the regional pattern. Within the Canberra region itself, though, the joints are internally consistent.

Central Australia Again, three regions were studied in Central Australia: (1) Near *Alice Springs*, where two outcrops were studied around Alice Springs (23°38′S, 133°56′E). The rocks were various types of metasediments, mainly schists. (2) In the *"Olgas"* (25°27′S, 131°13′E) which are a hill-complex 20 km S of Ayers Rock. Outcrops were found in the massive conglomerate which was formed from Proterozoic and Paleozoic sediments and folded during a Devo-

Table 2.7. Evaluations of Australasian joints, strike directions and earthquakes (P and T axes)

Location	No.	Max 1	Max 2	Angle	Bisectrices	
Australia						
South-East						
Canberra	64	151±14	55±10	84	13	103
NSW Coast	66	6±11	104±11	82	145	55
Snowy Mtns	69	4±19	88±09	84	47	132
S-Centr.Aust.	88	12±30	99±30	87	56	146
All Joints	287	5±11	93±09	88	49	139
Center						
Alice Spgs.	43	11±13	90±23	78	51	140
Mt. Olga	22	18±12	101±14	83	59	149
Macdonnell Rge	90	07±11	105±12	83	146	56
All Joints	155	10±08	101±09	89	145	55
Top End						
Mindli Beach	21	21±21	113±11	88	157	67
East Point	21	33±33	125±10	88	169	79
All Joints	42	27±16	119±08	89	162	73
New Zealand						
North Island						
Joints	63	179±14	84±14	85	42	132
Earthquakes T	27					138
South Island joints						
Christch.&Banks	81	171±14	84±10	87	128	38
Arthur's Pass	66	119±13	23±14	83	161	71
Picton	63	108±12	20±15	88	64	154
Molesworth	42	118±12	31±12	88	74	164
Omihi Val.	21	122±17	38±14	84	80	170
All South Island						
Joints	273	114±00	22±05	89	158	68
Earthquakes P, T	13				125	33
Fiji Islands						
Viti Levu	316	172±06	76±15	84	34	124
Yasawa Group	153	139±14	79±08	60	109	19
Mana Island	42	161±12	69±13	87	25	115
Vatulele Isl.	21	112±22	48±12	64	80	170
All Fiji	532	160±05	71±05	90	26	116

nian orogeny. (3) In the *Macdonnell Ranges* (24°S, 134°E), where four outcrops were investigated. These ranges consist generally of lower to middle Proterozoic rocks; gorges break through Precambrian (1000 Ma old) dolomite, limestone, shale and sandstone. The joint orientations were again *evaluated* according to the standard statistical proecedure; the results are listed in Table 2.7. There is once more a remarkable consistency between the orientations at the various locations.

The North   Finally, two outcrops (at Mindli Beach and at East Point) were studied in the vicinity of Darwin (12°25'S, 131°00'E). Both showed Cretaceous mudstone, limestones conglomerates and shales. The evaluations are given in Table 2.7.

Discussion   Considering the regional results one is struck by the basic consistency of the *joint strike directions* observed from the Victoria coast in the South to Darwin in the North, over a distance of about 3000 km; there is just a steady minor clockwise rotation of 23° from South to North (i. e. 6° from South to Center, 17° from Center to North). This concurs with the experience in Europe and North America, where the joint orientations are also basically the same over thousands of kilometers. *Stress determinations* from in-situ measurements and earthquake fault plane solutions show a picture of extreme complexity (Denham et al. 1979). The principal stress directions found in mines appear to vary around the clock; fault plane solutions of the few *earthquakes* that have occurred in Australia seem to indicate compression axes between 130° and 165°; which would fall into the range of one of the bisectrices of the joint orientations, pointing toward a possible geotectonic origin of the joints.

2.3.3.3
New Zealand

Introduction   New Zealand, lying roughly in the region from 34°S to 47°S and from 164°E to 179°E, consists of two main (and several smaller) islands: the North Island (114,687 km) and the South Island (150,460 km), separated by the 20 km-wide Cook Strait (cf. Fig. 2.8). A morphotectonic study of New Zealand

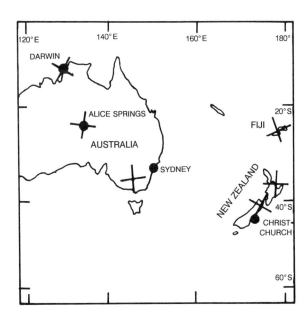

Fig. 2.8. Preferred joint strikes in Australasia (modified after Scheidegger 1995)

was reported in summary by Scheidegger (1995). The geology of New Zealand has been described in depth by Suggate (ed. 1978); summarized formulations have been given by Cumberland (1966) and Stevens (1975). In fact, the two main islands are geologically and tectonically quite different, and we shall deal with them separately.

North Island   The *morphology of most of the North Island* is hilly, its surface consisting of weak sedimentaries and marine, alluvial and volcanic accumulations. A volcanic plateau, or thermal zone, exists in the center of the North Island which contains the greatest expanse of Recent and Pleistocene volcanic materials in New Zealand. The underlying greywackes and the faulted structure of the basement are hidden underneath great ignimbrite sheets which were probably formed by the consolidation of hot rhyolitic fragments suspended in gas clouds from catastrophic Pleistocene eruptions (Walker, 1981). Hot springs/geysers and great active volcanoes, such as Tongariro and Ngauruhoe are common. Their lava flows have dammed up lakes like Lake Taupo and caused waterfalls such as Huka Falls below which the Waikato River forms a deep canyon in the softer material (Cumberland, 1966).

  Joint orientation measurements have been kindly contributed by Prof. R. Hantke of Zürich from two North Island locations: (1) Near Huka Falls in the vicinity of Weiraki, and (2) near Aratiatia 6 km downstream from Huka Falls, both on the banks of the Waikato River. The data were evaluated according to the usual statistical method; the results are shown in Tables 2.7. These results can be compared with the tectonics of the North Island which is viewed as extensional (Walcott, 1984, 1993) with an E-W maximum tension turning to NW-SE towards the east coast. Indeed, one of the bisectrices of the joint maxima has a trend-azimuth of N132E which fits together well with Walcott's views. Furthermore, these results also fit together with the fault plane solutions of earthquakes published by Evison and Webber (1986) who gave results in numerical form for the P(compressional)- and T(tensional)-axes in Central New Zealand. We have statistically evaluated those referring to the North Island and obtained $P \sim 334(154) \pm 42/32 \pm 18$; $T \sim 318(138) \pm 20/13 \pm 20$. The result for the P-axes is obviously not significant because of the steep plunge and the large error (more than 40°), but the mean T-direction agrees within a few degrees with one of the bisectrices found from the joint orientation measurements (Table 2.7): this supports the interpretation of joints as neotectonic signatures.

South Island   *Geologically*, three quarters of the South Island are covered with mountains. The Southern Alps form the spine of the island and reach their peak in Mt. Cook (3765 m). The main Alpine fault is a strike-slip fault with the sides having different throws, causing the relief; there is no nappe structure. Tectonically, it is generally assumed that the shear is the result of an E-W convergence of the Indian and the Pacific Plates (Allis 1981). Further north, several ranges of mountains occupy almost the whole width of the island which are separated by deep valleys which follow tectonic lines of transcurrent faults. These are all heavily folded and faulted and consist of lower Mesozoic sandstones and argillites. To the east in Marlborough the mountains descend to the sea, in Central Canterbury they give way to the lowlands of the Canterbury

Plains, which are interrupted by the deeply-eroded and sea-invaded twin mid-Pliocene volcanic domes of Banks Peninsula.

Joint orientation measurements. (1) In the vicinity of *Christchurch* (43°30′S, 172°38′E), the largest city of the South Island, joint orientations were measured at three locations on the road to a high ridge just E above. The ridge rises to nearly 1000 m above sea level ("Christchurch Summit"); it has quite steep slopes. Joint orientations were measured at one outcrop on the Banks Peninsula (see above). (2) The railway line from Christchurch to Greymouth across the Southern Alps has a stop at the pass (*Arthur's Pass*); in its vicinity it was possible to visit three outcrops. Complexly faulted geosynclinal and faulted geosynclinal greywackes and argillites made up the rugged terrain. The rocks contain Upper Triassic to Jurassic fossils. (3) Another railway line leads to *Picton*, at the head of the Queen Charlotte Sound. The latter forms a deep inlet in Carboniferous thin-bedded greywacke and argillites, interspersed with some rare volcanics. Joint orientations were measured at three locations at the NE side of the inlet. (4) A road leads inland from Picton through the Awatere and Acheron valleys south to Blenheim and then via *Molesworth* Cattle Station to Hanmer Springs: The original forest (Podocarpus-Nothofagus) had been completely clear-felled for making sheep-pastures, but this caused the valley sides to become barren with unstable scree slopes. Joint orientations were measured at two locations. (5) Finally, the vicinity of the Glynn Wye- and Hope-Faults were visited, and 21 joint orientations could be measured on the sides of the *Omihi Valley* in loamy-sandy clay. The joint orientations at the mentioned outcrops were split into the five groups corresponding to the five regions considered above and evaluated according to the usual statistical method; first all the groups individually and then all values from the South Island were measured (Table 2.7).

Interpretation. As noted in the section on geology, it is generally assumed that the tectonics of the South Island is the result of a compression (in contrast to the North Island where it is the result of an extension), inasmuch as the boundary between Indian and Pacific tectonic plates lies to the west of the South Island, but to the east of the North Island (Allis 1981). In the South Island, this compression was postulated by Allis (1981) to be in a general East-West direction. The study of Evison and Webber (1986) also contains 13 earthquake fault plane solutions falling into the neighborhood of the South Island, albeit only at its North. Their directions of the *P*- and *T*-axes were equally subjected to a statistical analysis: The means of the 13 earthquakes in question were for *P*: 125 ± 14/ ± 06, and for *T*: 213(= 33) ± 24/27 ± 22. The result for *T* is obviously not significant, but that for the maximum compression indicates a NW-SE direction, which concurs more or less with the view of a general E-W compression in the region, as well as with one of the bisectrices of the joint maxima lying between 128° and 170°. At any rate, the earthquake fault plane solutions clearly support the hypothesis of a stress reversal between the North and South Islands of New Zealand.

2.3.3.4
Fiji

Introduction   The *Fiji* Archipelage encompasses more than 300 islands, centered around 18°50'S and 175°E at the NE corner of the Australasian Plate: it lies between the N end of the Tonga trench and the S end of the Vanuatu trench. The crustal movements in the *Fiji archipelago* have been quite complex; the Pacific plate is supposed to move W with respect to the Australasian plate. It is criss crossed by many faults (Rodda, 1994).

Joint orientation measurements   Joint orientations were measured as follows.

(1) On *Viti Levu*. This island, centered around 18°S and 178°E, is the largest (with 10,384 km^2) and geologically oldest island of the Fiji group; the oldest rocks are exposed S of Nandi on the north coast and consist of andesitic volcanics as well as of limestone of upper Eocene age. The basement rocks are covered by thick sequences of lava flows and volcaniclastic rocks, ranging from tholeiitic basalt to dacite. A major orogenic period occurred in the middle Miocene, in which also widespread sedimentation took place. In the Miocene, volcanism occurred which continued up to the early Pliocene (Rodda, 1994). Joint orientation measurements were made by the author at six outcrops around the coast; Mr. Peter Rodda, Principal Geologist of the Department of Mineral Resources in Suva, kindly contributed measurements from the Namosi area in the center of the island.

(2) 163 joint orientation measurements were also contributed by Mr. P. Rodda (personal comunication) on *the Yasawa islands*. These are centered around 17°S and 177°E; they consist of basaltic and dacitic volcanics and show evidence of strong compression. They were formed by late Miocene eruptions (Rodda 1994).

(3) To the S of these islands lies the Mamanuca group with *Mana Island* where the author measured 42 joint orientations.

(4) Finally, Dr. Patrick Nunn, Reader in Geography at the University of the South Pacific in Suva, kindly contributed 21 joint orientations from *Vatulele Island*, situated 60 km to the S of Viti Levu: Vatulele is a low-lying limestone island created by Oligocene-Miocene submarine volcanism followed by limestone accumulation and subsequent uplift (Nunn 1988, 1994).

Interpretation   The joint orientation measurements in the four locations mentioned were statistically evaluated by the usual method; the results are summarized in Table 2.7. An inspection of the values shows that one maximum (Max. 2) between 69 and 79° is present in all regions except in Vatulele Island, the second maximum (Max. 1 in Table 2.7) varies a little more (139–172°). For all of Fiji, the bisectrices (presumably the principal stress directions) have trends of 26° and 116° with an angle between the conjugate joint being exactly 90°. One would probably take the 116°-stress trend as *P*-direction for the whole of Fiji. Vatulele Island may well be an exception, inasmuch as Nunn (1988) noted that the "joints" on that island are really faults trending normal to the shore line.

2.3.4
South America

2.3.4.1
General Morphology/Geology

South America is the southern part of the American continent. As noted earlier, it has a different geological history as North America, inasmuch as the latter developed from the Laurasian, South America from Gondwana supercontinent. In the west of South America there is a large chain of high recent mountains, the Andes, which are partly split into several subparallel chains; they border the Pacific Ocean. Adjacent to the east there are regions of plains (the Pampas and the Selva), consisting of sedimentary areas surrounding and covering a Precambrian shield. The plains/shield abut directly on the Atlantic Ocean in the east and find a direct continuation in Africa. In the NE of South America there is an offshoot of the Andes in the form of the Cordillera de la Costa of Venezuela which continues into Trinidad. The Lesser Antilles are situated to the north of South America; some of them actually belong morphologically to the latter (Aruba, Bonaire, Curaçao, Margarita etc.), but others are assigned to a special "Caribbean Plate"; the division is not always clear (cf. Trinidad). The Lesser Antilles will be treated along with continental South America in the present section.

2.3.4.2
Northern South America

Scheidegger and Schubert (1989) have summarized joint orientation measurements from outcrops on the coastal plains of Venezuela (Maracaibo), along the Venezuelan and Colombian Cordilleras, in the Llanos of Colombia and in the Guayana Shield of Venezuela.

Accordingly, joint orientations were measured in the *vicinity of Maracaibo:* three outcrops on the lake shore (Quaternary sands, silts etc.), four on the island of Toas (Paleozoic granites and early Cretaceaous sands) and three in the vicinity of the coal mines on the Guasare River (coal-bearing Paleozoic sandstones, limestones and shales). In the Venezuelan *Coast Range,* the outcrop-rocks consisted of quartz–mica schists, mesozoic(?) granitic gneisses; at one coastal location, of marine sediments (details in Schubert and Scheidegger 1986). In the *Eastern Venezuelan Andes* measurements were made along the Boconó Fault; the rocks encountered consisted of Paleozoic phyllites and limestones as well as of Jurassic/Cretacous shales. Many joint orientations were measured around Mérida in the *Western Venzuelan Andes.* In general, the rocks consisted of Precambrian schists, quartzites and granites, as well as of a few Tertiary sandstones. In the *Northern Colombian Andes* joints were measured mainly in Mesozoic sediments, in a few Cambrian schists, Precambrian gneisses and Recent andesites (details in Mojica and Scheidegger 1981). The three southern Cordilleras of Colombia were investigated between Florencia, Buenaventura, Manizales and Bogotá. The majority of outcrops consisted of Mesozoic and

Fig. 2.9. Joint strike orientations in investigated regions of South America: *1* Maracaibo; *2* Venezuelan Coast Range; *3* Venezuelan Eastern Andes; *4* Venezuelan Western Andes; *5* Colombian Northern Andes; *6* Colombian Southern Andes; *7* Colombian Llanos; *8* Venezuelan Shield; *9* Curaçao; *A* Lesser Antilles; *U* Fiordo de la Ultima Esperanza; *D* Cordillera Darwin; *E* Isla de los Estados and other South Argentinian Islands; *B* Regions South of the Beagle Channel

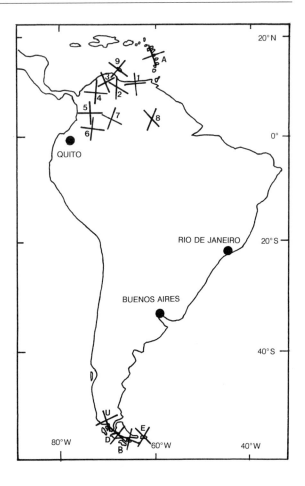

Tertiary sedimentaries; a few of Recent volcanics (details again in Mojica and Scheidegger 1981). Turning towards the interior, joint orientations were measured east of Bogotá in the *Colombian Llanos*, from the foot of the Andes to Puerto Lopez. The rocks near the Cordillera were Ordovician sandstones and limestones, changing to Quaternary sandstones in the Llanos proper. Finally Dr. Schubert of Caracas measured joints at 18 outcrops in the Guyana Shield of Venezuela (Briceno and Schubert 1985). The rocks encountered in the latter were Precambrian quartzites.

The results of the statistical evaluations of the joint orientation data are shown in Table 2.8 and in Fig. 2.9. An inspection of these results shows first of all that the joint/stress orientation pattern in the coastal area of Maracaibo correlates neither with that in the adjacent Andes, nor with that in Curaçao (see Sect. 2.3.4.3). It is well known that this region is crisscrossed by a number of active faults (Oca, Boconó, and Santa Marta Faults; for details see the cited paper of Scheidegger and Schubert (1989)), so that there is no uniform stress state in it.

Table 2.8. Orientations of strikes of joints/faults and trends of earthquake *P*-axes in South America and the Carribean Islands

Location	No.	Max 1	Max 2	Angle	Bisectrices	
Northern South America						
Coast						
Maracaibo	231	156±08	59±10	83	17	107
Venezuelan Cordillera						
Joints						
Coast Range	635	172±06	88±09	87	130	40
E Andes	233	117±30	4±30	67	151	61
W Andes	429	188±07	95±05	88	52	142
Earthqu.P	6					117
Colombian Cordillera						
Joints						
N Andes	1159	91±04	179±05	88	134	25
S Andes	1784	188±06	100±05	88	144	54
Earthqu. *P*	1				135	
Interior						
Col.Llanos	339	117±11	24±15	88	161	71
Ven.Shield	796	130±06	22±02	72	166	76
Curaçao	136	124±12	34±12	90	170	80
Lesser Antilles						
St.Maarten	439	145±08	60±07	85	103	13
St.Barthélémy	293	143±06	53±08	90	98	8
Antigua	254	173±13	85±10	88	129	39
St.Kitts	144	163±14	78±09	83	119	29
Montserrat	179	170±12	74±13	84	32	122
Guadeloupe						
Joints	1557	166±11	76±09	88	32	122
Fractures	119	163±18	70±12	85	27	116
Dominica	111	108±17	19±16	89	63	154
Martinique	722	162±07	75±08	87	118	28
St.Lucia	45	146±17	77±40	80	115	20
St.Vincent	93	13±18	78±32	83	40	137
Barbados	261	115±14	33±12	87	74	164
All Less. Ant.						
Joints	11*	161±23	72±12	89	117	27
Earthqu. *P* ax.					123	
Southern South America						
Joints						
Ultima Esper.	148	158±09	72±04	86	25	115
Cord. Darwin	251	129±11	49±04	80	89	179
S Argent.Is.	90	131±00	29±00	78	80	170
S of Beagle	127	111±03	28±00	83	68	159
All regions	4*	132±22	43±23	88	89	178
Earthqu.P ax.	3				66	

*regions containing many measurements each

Along the coast range in Venezuela and along the Andes (regarding the geology of the latter, see the review by Zeil 1979) in Venezuela and in Colombia, however, there is a very uniform stress state with an azimuth of one of the bisectrices (presumably the maximum compressive [P] stress) around 140° (viz.130°–151°). The interpretation that the NW-SE direction is consistently the P direction in this area is supported by a comparison with fault plane solutions of earthquakes: in the case of the Popayan earthquake of 31 March 1983 the P-axis was found to trend to 135° (Mayer et al. 1986); in Venezuela, 6 fault plane solutions are available (Pennington 1981) for which the statistical mean trend was calculated to have an orientation of 117°/4° (Scheidegger and Schubert 1989), i.e. in the SE-quadrant like one of the joint-bisectrices.

In the interior regions represented by the Colombian Llanos and the Venezuelan Guayana Shield, NW-SE is still the quadrant containing one of the bisectrices, presumably the P direction, but the azimuth of the latter is turned somewhat (to 162°) compared to its trend in the north. The reason for this may be that the boundary between the S American and Caribbean Plates trending EW is supposed to be a shear zone: Thus the principal stress direction is forced by the boundary conditions to turn so as to form an acute angle with this shear line.

2.3.4.3
Caribbean Islands

The situation of the Caribbean Islands is ambivalent: Whereas Curaçao, Bonaire etc. seem to be a continuation of the South American land mass, and the ridges on Trinidad clearly a continuation of the Cordillera de la Costa of Venezuela, the *Leeward and Windward Islands* are on the rim of a supposed "Caribbean" plate. Moreover, these islands belong to two morphologically distinct arcs: an inner (western) "volcanic" and an outer (eastern) "calcareous" arc. Of the islands investigated, the calcareous arc comprises St. Maarten, St. Bartélémy, Antigua and Barbados; the volcanic arc St. Kitts, Monserrat, Dominica, Martinique, Ste. Lucie and St. Vincent. Guadeloupe straddles the two arcs, with one part (Grande Terre) being calcareous, the other (Basse Terre) volcanic. The volcanic arc stretches about 900 km from N to S, containing basalts, andesites and dacites with ages ranging from the late Miocene or early Pliocene to the present (eruption of Montagne Pelée on Martinique in 1902). The calcareous arc contains, in fact, volcanic centers as well, which were active before the Miocene (Weyl 1966), but, after a period of subsidence and erosion, reef-limestones were deposited, covering the volcanics.

Measurements of joint orientations have been reported by Bonneton and Scheidegger (1981) from the *Leeward and Windward Islands* between St. Maarten and Barbados; Curaçao has been investigated and treated separately by Schubert and Scheidegger (1986). The results of the evaluations of the data of the 11 Leeward and Windward Islands mentioned are shown in Table 2.8. There is a general orientation of one of the bisectrices (presumably the P direction) in the NW-SE quadrant; this becomes particularly evident if a combined evaluation of all the islands is made. It is interesting to note that this fits with the directions of the volcanic fractures around the Soufrière on

Guadeloupe (163 ± 18, 70 ± 12; cf. Table 2.8), which have been determined by Bonneton and Scheidegger (1982), as well as with the maximum compression direction obtained from earthquake fault plane solutions in the area; the latter yields $P \sim 123 \pm 34/10 \pm 32$ (Bonneton and Scheidegger 1981). The bisectrices (presumably principal stress directions) of the joint orientations in the Leeward and Windward islands fit neither into a general pattern which would be related to those of North or South America; these islands may indeed belong to a separate "Caribbean" plate belonging neither to Laurasia nor to Gondwanaland.

Curaçao has been treated separately by Schubert and Scheidegger (1986); it is one of several Caribbean Islands situated directly offshore to the South American Continent [like Aruba, Bonaire, and Margarita]. The basement rocks of Curaçao belong to a folded series of submarine basic lavas and tuffs; on it rest unconformably Upper Cretaceous rocks composed of chert-rich sediments, conglomerates, limestones and tuffs (Weyl 1966). Its joint orientation values (Table 2.8) and therewith its tectonic stress field correspond rather surprisingly to those found in southern Mexico (cf. Table 2.3), not to those found in South America or in the Antilles.

2.3.4.4
Southern South America

Joint orientation measurements were made in some key areas of the Scotia Plate on and around the southern tip of the American Continent, viz. in the Ultima Esperanza and the Cordillera region of southern Chile, as well as on the islands south of the Beagle Channel in southern Argentina and Chile (Scheidegger 1990). All these regions belong to the "Coastal Cordillera" of southern South America (Zeil 1979); details on the areas visited were described by Dalziel (1989).

The northernmost region visited was the *Ultima Esperanza Region*, in which layer sequences from Jurassic volcanic rocks (Tobifera formation; volcaniclastic sandstones positioned above well-bedded tuffs), to sedimentary sequences derived directly or indirectly therefrom, intruded by basalts (in the Miocene as well as in the Pleio-Pleistocene) and granites (in the Miocene) were encountered. Joint orientations were measured at seven outcrops.

Further south rise the southernmost Andes which are called "*Cordillera Darwin*", representing the main E-W trending range in the area. According to Dalziel (1989), it is formed by a high in the pre-late Jurassic basement of the southermost Andes which is seen in a structural window of the cover-rocks. The latter are mainly Jurassic^{a}Cretaceous volcanic breccias (Tobifera formation, as above), overlain originally by rocks formed in a marginal basin of the southern Andes, injected at various times by intrusives such as the Darwin Granite in the Jurassic and the Beagle Granite in the Cretacous. Joints were measured at four outcrop groups.

The *Southern Argentine Islands* have to be regarded as a continuation of the main Andes chain. Joint orienations were measured on the Isla de los Estados and in the Argentinian part of the Isla Grande (Tierra del Fuego).

Finally, the geology of the *islands south of the Beagle Channel* conforms to that of the Cordillera Darwin. Of particular interest are exposed ophiolites as well as Darwin granites and Patagonian batholiths intruded on some of the islands visited.

The data were evaluated statistically; the results are summarized in Table 2.8; drawings of the preferred joint strikes are shown in Fig. 2.9. It is observable that the majority of the locations give essentially E-W as one of the bisectrices (presumably the maximum-compression (P) direction); an E-W compression fits with the idea of an active EW subduction from the Pacific side toward the Scotia Plate. This interpretation is supported by the directions of the P-axes in 3 fault-plane solutions of earthquakes by Pelayo and Wiens (as reported in Dalziel, 1989), which have a general direction of more or less E-W ("average" 66°). Thus, a conformity of the observed joint orientations with standard plate tectonic ideas is found.

2.3.5
Antarctic and Subantarctic Regions

2.3.5.1
General Morphology/Geology

The Antarctic continent is a mostly icecovered land mass surrounding the South Pole; only the coasts afford access to the country-rocks. It is divided into East and West Antarctica; East Antarctica is a Precambrian cratonic shield separated by the Transantarctic Escarpment from West Antarctica; the latter juts northward toward South America forming the Antarctic Peninsula. It is paralleled on the NW-side at a distance of roughly 120 km by the South Shetland Islands. Joint orientation measurements were made in West Antarctica on and around the Antarctic Peninsula: on the Peninsula itself and on the mentioned South Shetland Islands (cf. Scheidegger 1990). Since there are no roads, access to the outcrops has generally to be gained from zodiac boats and on foot from their landing places.

2.3.5.2
South Shetland Islands

Physiography   The South Shetland Islands reach from Clarence Island (54°W, 61°S) to Smith Island (62°30′W, 63°S); they lie on two axes which are about 20 km apart: a northern, non-volcanic axis with Elephant, King George, Livingston, Gibbs (and some smaller) Islands, and a basaltic volcanic southern axis with Bridgeman, Deception and the Low Islands; the non-volcanic northern islands are geologically older than the stll presently active volcanic southern islands. The entire archipelago is separated from the Antarctic Peninsula by the 120 km-wide Bransfield Strait, which started opening up about 3 Ma ago (Dalziel 1989).

General Geology The South Shetland Islands represent a magmatic arc on the west side of the Antarctic Peninsula which has been active since Silurian times up to the present (cf. the large eruption on Deception Island in 1970). Most of the non-volcanic islands consist of subduction-complex rocks: ductile melanges, metacherts, blue- and greenschists and intermediate amphiboles that may date back to the Permian times. Some parts of the islands though, may be part of the Antarctic Peninsula cut off by the tectonic opening up of Bransfield Strait starting 3 Ma ago which is still ongoing today.

Elephant Island Group The easternmost of the South Shetland Islands are referred to as the "Elephant Group" ("elephant" refers to sea-elephants, Mirounga leonina, not to "regular" elephants such as Loxodonta africana!). Joint orientations were measured on Seal Island (55°24'W, 61°02'S) and Elephant Island (54°45'W 61°07'S). Seal Island consists of siltstones (unmetamorphosed) and conglomerates containing volcanic greywacke, micritic limestone and chert pebbles. Elephant Island is composed of metamorphic rocks interpreted as part of an uplifted subduction complex (green and grey phyllite, volcanic breccia and thin marble layers).

King George Island This, the largest island of the archipelago, has a military all-weather airstrip at the Chilean Marsh Base (58°W, 62°S). Its highest elevation is 600 m. There are many sheltered bays on the south side. It is largely composed of easily eroded late Cretaceous to Cenozoic (mainly basaltic) volcanic arc rocks. Longitudinal faults delineate several tectonic blocks. Motions along the faults began at the Paleocene-Eocene boundary; volcanism is documented up to the Quaternary. Three outcrops were investigated.

Livingston Island This is the second-largest island (60°W, 62°30'S) of the archipelago with peaks up to 400 m. The body of the island consists of a Mesozoic/Paleocene melange, but its south side seems to be a torn-off part of the Antarctic Peninsula. Sheltered bays along the south coast afforded access to three outcrops, consisting of Permo-Triassic forearc sediments.

Deception Island A special morphotectonic study was made of Deception Island (60°40'W, 62°58'S), which is the only island visited on the *volcanic southern axis* of the South Shetland Islands. It is circular, about 12 km in diameter; the rim of the circle is formed by volcanic vents; the central depression is a caldera filled with sea water; the latest eruption on Deception Island took place in 1970. Normal faults form a grid on the island (Dalziel 1989, p. 151). The morphology described has been considered as "epigenetically" (i. e. as essentially purely volcanically and glacially) created, but we contend that there has also been an endogenic co-design. In order to show this, the orientation structures of the 21 joints measured around the central caldera and the trends of the faults noted by Dalziel were subject to the usual statistical analyses. The resulting numerical values are shown in Table 2.9: It will be noted that the first maxima of joint and fault sets correlate within their error limits; the second maxima also correlate (almost) within these limits, but the latter being very large and the second joint maximum being very diffuse, this is perhaps not too significant. At any rate, the second joint maximum correlates very well with

Table 2.9. Evaluation of joints/faults/earthquakes in West Antarctica Strike/Trend Directions

Location	No.	Max 1	Max 2	Angle	Bisectrices	
South Shetland Islands						
Elephant (& Seal)	59	92±18	22±13	70	57	147
King George	66	95±00	178±03	82	147	46
Livingston	63	105±16	8±13	83	146	56
Deception						
Joints	21	105±11	17±30	88	61	151
Faults	43	116±06	48±00	68	82	172
All S.Shetlands						
Joints	209	93±12	5±15	88	40	139
Earthquake-*T* axes	7					140
Antarctic Peninsula						
Bahía Esperanza	42	130±16	25±13	75	168	78
Anvers Island	63	109±09	20±10	80	60	150
Lemaire Channel	87	138±12	32±19	74	174	84
All Peninsula	192	123±08	24±00	81	163	73

the corresponding one found for the South Shetland Island joints as a whole (Table 2.9). This shows that many features on Deception Island have been co-designed by neotectonic processes.

2.3.5.3
Antarctic Peninsula

General Physiography   West Antarctica consists of a group of younger continental blocks about 30 km thick from which the Antarctic Peninsula juts northward from a latitude of about 75°S to 63°14′S between longitudes of about 57°W and 70°W. It is slightly curved northward towards the east; its east side is bordered by the Weddell Sea, its west side by the Pacific part of the Antarctic Ocean. It is mostly ice-covered; the West Coast is quite broken up by channels parallel to the coast (straits) which have many (in the Antarctic summer) ice-free bays and coves where landings can be effected and outcrops can be investigated. The islands west of the channels are rugged and consist mainly of volcanics.

Geology   According to Dalziel (1989), the Antarctic Peninsula developed as a magmatic arc above a subduction zone in the Pacific starting with the initial breakup of Gondwanaland in the Jurassic. A large part of the Peninsula consists thus of early Mesozoic forearc deposits, mainly turbidites (Trinity formation) deposited on and around late Paleozoic and Mesozoic plutons; the Weddell Sea represents a back-arc basin. There seem to be no deformations on the Peninsula that could be connected with the Andean orogeny, except for a "final" uplift in the mid-Cretaceous time. Granitic rocks were also variously intruded from the late Cretaceous to early Tertiary times. A thick succession of volcanics

(Trinity volcanics) can be observed on the E side of several channels (Dalziel 1989).

Bahía Esperanza (Hope Bay)   This is the location of an Argentine Station at the tip of the Antarctic Peninsula (57°00'W, 63°29'S). The shore rocks consist of Permo-Triassic Trinity volcanics; walking further up toward the central ridge of the Peninsula, one crosses an unconformity and encounters clastic terrestrial Mesozoic backarc rocks (measurements at two outcrops).

Anvers Island area   This is an island on the west side of one of the channels (Neumayr Channel) breaking up the west coast of the Peninsula mentioned earlier. It is the location of an American Antarctic station. In the vicinity of the latter, the bedrock is primary batholithic tonalite at the station itself and granite-diorite at an exposed point at the shore. On an adjacent island, andesitic lava cut by a dyke was encountered. Joint orientation measurements were made at three locations.

Lemaire Channel   Finally, another coast-parallel channel (Lemaire channel) was followed southward to latitude 65°10'S, along which landings at four locations were possible: the outcrops showed hornblende-diorite granite and Jurassic (fossil bracchiopodes!) volcanic tuffites.

2.3.5.4
Interpretation

As noted above, joint orientations were measured on various South Shetland Islands and on the Antarctic Pensinsula; the results of the usual statistical evaluations are shown in Table 2.9 and the preferred joint strikes in Fig. 2.10.

The preferred orientations of the joints for the various regions show that there is a remarkable uniformity over large (100 km or so) distances that are comparable to the dimensions of the tectonic plates. Taking the bisectrices of the joint sets as representing principal stress directions, it becomes clear

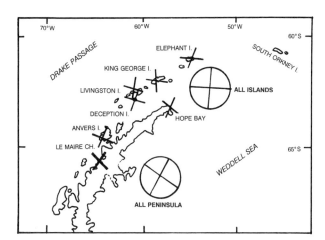

Fig. 2.10. Preferred joint strikes in Antarctica

that there is a stress (bisectrix) direction in the NW-SE quadrant (azimuth N139°E) in the South Shetland Islands, which is exactly normal to the trend of Bransfield Strait; if it is interpreted as the *maximum stress relief* ("tension") direction, it would correspond to the opening up of that Strait during the last 3 Ma: the joints are simply aligned in the shear planes of the geodynamic strain. This interpretation is supported by the results of a series of earthquake fault plane solutions made by Pelayo and Wiens and reported by Dalziel (1989), seven of which concern the South Shetland Islands. The pressure (P) axes of these earthquake scatter widely, but the "mean" for the T-axes, statistically calculated by the Kohlbeck–Scheidegger (1977) algorithm, yields $140° \pm 22°$ (see Table 2.9). This fits exellently together with the interpretation of a stress-relief occurring across Bransfield Strait.

2.4
The Oceans

2.4.1
General Physiography/Geology

By far the largest part of the Earth's surface is covered by oceans. Customarily, one considers three "world-oceans", (1) the Atlantic Ocean between the continents of Eurasia and Africa on the one hand and the Americas on the other, (2) the Indian Ocean between Africa, Persia and Peninsular India and (3) the Pacific Ocean covering almost a hemisphere around the 180°-Meridian. These oceans are all more or less "open" towards the south, where they are connected by the "Southern Ocean" which surrounds the Antarctic Continent. Since our discussion of "global morphotectonics" is heavily based on joint orientation measurements, we will mainly confine ourselves to islands.

The most extended morphological elements of the ocean-bottoms are the "abyssal plains", large areas of somewhat hilly (300 m or so) relief whose mean depth of submersion is 5–6 km below the sea surface. The abyssal plains are broken up by mid-oceanic ridges which form a worldwide system as shown in Fig. 2.11. The ridges are much higher and broader than land-mountains, they reach from the depth of the abyssal plains almost to the sea surface, sometimes above it (Iceland, Ascension). At their base they are 1500–2000 km wide; at their crest, there is often a rift in which volcanic activity occurs. The plate-tectonic view is that the ocean basins grew by spreading from the ridges connected with the volcanic activity at their crest-rifts (cf. Sect. 1.6.2) which would engender the drift of tectonic plates as well as the genesis of lines of volcanic islands by the drift of the plates over and across mantle hot spots (Dietz 1961). However, the ocean floors are generally criss crossed by what almost appears as a grid of conjugate lineaments (ridges and fractures) which do not point to a formation by plate drift, but rather to a formation as shearing lines in a stress field. Such lines simply do not fit any idea of a plate drifting over a hot spots at all (cf. Sect. 1.6.3).

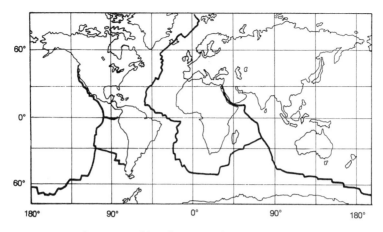

Fig. 2.11. World-wide system of mid-ocean ridges

2.4.2
The Atlantic Ocean

2.4.2.1
Morphology/Geology

The Atlantic Ocean is supposed to be the consequence of the separation of the erstwhile supercontinents of Laurasia (Europe and North America), and Gond-wanaland (Africa and South America) around Cretaceous times (Sect. 1.6.2). The fact that there is a rift zone ("Mid-Atlantic Rift") running NS in the middle of the Atlantic ocean actually concurs with such an interpretation. In the Atlantic we have studied the Macaronesian (Scheidegger 2002b) and Bermudan (Scheidegger 1985c) archipelagoes (cf. Fig. 2.12).

2.4.2.2
Macaronesia

Geography "Macaronesia" ("The Blessed Islands") encompasses the islands lying in the Atlantic Ocean between 10°N and 40°N latitude. The southernmost of these are the Cape Verde Islands, followed to the N by the Canary Islands, the Madeira Islands and the Azores Islands (cf. Fig. 2.12). The author (Scheidegger 2002b) has made a study of these islands to which the reader is referred for details; we shall summarize the results.

General morphology/geology All islands show a volcanic morphology, the volcanism starting in the Miocene and lasting up to the present. The volcanic material is mainly basaltic, but in spots also andesitic/trachytic; it lies on underlying older sediments (upper Jurassic; Storetvedt 1997, p. 181) which are intercalated with recent sediments (Mitchell-Thomé 1976). The basaltic lavas would conform to to the view that mid-oceanic islands represent a basaltic

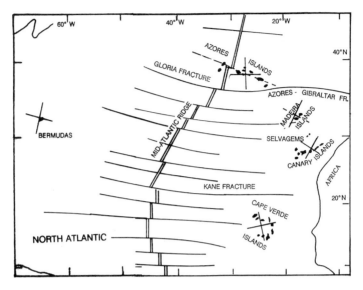

Fig. 2.12. Islands in the (North-) Atlantic Ocean. Cape Verde Islands (anticlockwise from South): Brava, Fogo, Santiago, Maio, Boavista, Sal, Sao Nicolao, Sao Vicente, Santo Antao; Canarias (from west to east): western group Hierro, La Palma, Gomera, Tenerife, Gran Canaria, eastern group Fuerteventura and Lanzarote; Madeira Archipelago (from S to N): Madeira, Porto Santo (Desertas not shown); Azores (from W to E): Flores and Corvo W of the mid-Atlantic Ridge, then Faial/Pico, S. George, Graciosa, Terceira (central group), finally in the east: S. Miguel and Santa Maria. The main islands of the Bermudas are shown as *one* complex, since they are connected by bridges. (Modified after Scheidegger 2002b)

crust, but fits neither the presence of andesitic/trachytic material nor the presence of underlying older sediments.

Cape Verde Islands   The Cape Verde Archipelago is located in the shape of a horse-shoe between ca. 14°N and 17° N latitude and 20°W and 25°W longitude; it contains 15 islands (see map in Fig. 2.12) and comprises the oldest rocks of Macaronesia. The various islands show different stages of geomorphic development (Mitchell-Thome 1976). Major volcanism occurred in the Neogene (Stillman et al., 1982) and is continuing up to the present (latest eruption on Fogo on 2 April 1995). Joint orientations and stream directions were measured on six islands (S. Antao, S. Vicente, Sal, Maio, Santiago and Fogo); ridge direction measurements on two islands (S. Antao, Santiago). The preferred joint strike directions approach the directions of the mid-Atlantic ridge and the lineaments around it (for details, see Scheidegger 2002b). The same is more or less true for the orientations of the stream-links and the ridges (Table 2.10). Thus one can conclude that all these features may be due to the same cause: the action of the neotectonic stress field.

Canary Islands   The Canary Archipelago, North and East of the Cape Verde Archipelago, contains seven main islands as shown in Fig. 2.12. It stretches from ca 27°N to 29°N and from 13°W to 18°W into the Atlantic Ocean (Medwenitsch 1970). Its eastern islands are aligned along a line trending N32°E (Fig. 2.12), its western islands are mountainous and and are aligned along two lines, one trending N65°E and the other N112°E and crossing each other at the center of Tenerife (Martinez de Pison and Quirantes 1994). Most major islands, viz. Lanzarote, Fuerteventura, Gran Canaria, Tenerife and La Gomera, were visited and studied by the author (details in Scheidegger 2002b).

The Canary islands are mainly volcanic; the volcanism has occurred in various episodes starting in the late Cretaceous (Meco and Stearns 1981; Le Bas et al. 1986; Martinez de Pison and Quirantes 1994; Rothe 1996; Scarth and Tanguy 2001) and was mainly basaltic, except for Tenerife, where the most recent eruptions have been acidic (trachytic-syenitic; cf. Mitchell-Thome, 1976; Rothe, 1996). The volcanic rocks overlie unconformably an older pre-volcanic (Cretaceous) "basal complex" which consists of *non-volcanic* sediments.

Joint- and stream-orientations were measured on five Canarian islands, ridge orientations on one. It has been found that the Max-1 directions between various features (joints, streams, ridges), agree generally with each other almost on every individual island, at least with regard to one maximum, although their azimuths vary from 110° to 142° between the individual islands. For Lanzarote and Fuerteventura the results are even almost identical; they are "sister islands" lying on the same submarine bank (cf. Fig. 2.12); the strikes of one of the joint maxima line up with the trend of the submarine bank (32°) on which they lie (details in Scheidegger 2002b). Combined for the entire archipelago one obtains for the Max-1 strike/trends of N121°E for the joints and N127°E for the stream-links for the Max-1 directions (Table 2.10). The Max-2 directions for the entire arichipelago are very close to each other (30° for the joints and 32° for the streams), which suggests a genetic relationship. If the bisectrices of "conjugate" morphotectonic direction-maxima are taken as the principal far-field tectonic stress directions 166°–169° and correspondingly 76°–79°, then the azimuths agree closely with the principal stress orientations obtained from a fault plane solution of the earthquake that occurred on 9 May 1989 between Tenerife and Gran Canaria (Mezcua et al. 1992): $P = 166°$ and $T = 75°$: this indicvates that valleys, joints and earthquakes may all have been caused by the same tectonic mechanism.

An *interpretation* of the above correlations in terms of "plate tectonics" has been suggested by Martinez de Pison and Quirantes (1994): the layout of the Canaries is a consequence of an anticlockwise rotation of Africa and the surrounding crust. This would be due to the ongoing W-E spreading Atlantic Ocean floor exerting an E-directed drag on the SW-trending coast, producing an anticlockwise torque causing a series of NE-SW and NW-SE running fractures.

Madeira Archipelago   The Madeira Archipelago stretches from about 32°N to 33°N and from 16°W to 17°W. Its main islands are Madeira proper (by far the largest) and Porto Santo (cf. Fig. 2.12). Both islands are basically of basaltic volcanic origin which started in the Miocene and died out long ago. The volcanics are mainly basaltic rocks with a few interspersed acidic trachytes

Table 2.10. Atlantic Ocean Islands. Collated from Scheidegger (1995, 2002b)[a]

Location	No.	Max 1	Max 2	Angle	Bisectrices	
Cape Verde Archipelago						
S. Antao						
Joints	146	166±05	77±07	89	121	31
Links	116	152±03	49±07	77	100	10
Ridges	18		81±17			
S. Vicente						
Joints	183	173±05	80±06	87	127	37
Links	42	144±07	52±07	87	98	8
Sal						
Joints	214	177±00	87±07	90	132	42
Links	19		74±13			
Maio						
Joints	173	1±16	80±00	79	131	41
Links	42	117±24	22±05	85	159	69
Santiago						
Joints	197	154±08	69±00	84	111	21
Links	126	163±10	52±05	68	108	18
Ridges	34	167±04	97±16	70	132	42
Fogo						
Joints	194	168±10	65±08	88	121	31
Links	107	175±00	69±00	74	122	32
Cape Verde All						
Joints	1107	168±00	78±00	89	123	33
Links	452	156±00	56±00	80	106	16
Ridges	52	169±04	90±10	78	130	40
Canary Archipelago						
Lanzarote						
Joints	88	120±10	33±09	87	167	77
Links	98	126±08				
Fuerteventura						
Joints	377	126±05	30±05	85	168	78
Links	534	117±03				
Ridges	60		56±06			
Gran Canaria						
Joints	54	132±19				
Links	562		5±05			
Tenerife						
Joints	259	140±00	77±00	64	108	18
Links	339	142±07				
Gomera						
Joints	80	110±15	14±13	83	152	62
Links	176	132±12	46±04	86	171	81

[a] Macaronesian results reprinted by permission of 2 Sep. 2003 from Scheidegger (2002b), Tables 1–4 © 2002 Elsevier Science B.V.

Table 2.10. (continued)

Location	No.	Max 1	Max 2	Angle	Bisectrices	
All Canary Islands						
Joints	858	121±08	30±09	89	166	76
Links	1709	127±00	32±00	84	169	79
Ridges	60		56±06			
Earthquake	1				166	75
Madeira Archipelago						
Madeira						
Joints	162	109±11	21±08	89	155	65
River links	157	143±00	22±06	59	172	82
Porto Santo						
Joints	63	122±10	29±08	87	165	75
River links	45	130±11				
All Madeira Archip.						
Joints	225	116±10	26±08	89	161	71
River links	202	128±03	18±00	70	163	73
Azores Achipelago						
Sao Miguel						
Joints	126	177±07	86±09	88	131	41
River links	322	173±01	48±00	55	110	20
Ridges	33		90±07			
Terceira						
Joints	312	172±04	78±00	86	125	35
River links	184	4±01	56±00	52	120	30
Sao Jorge						
Joints	142	18±11	105±09	87	151	61
River links	88	20±02	131±09	68	165	75
Ridges	43		115±04			
Flores						
Joints	180	23±10	114±02	89	158	68
River links	118		119±07			
Corvo						
Joints	105	170±13	77±12	86	123	34
River links	26	149±16	72±06	76	111	21
Azores all						
Joints	865	3±01	90±01	86	137	47
River links	738	11±00				
Ridges	76		102±04			
Bermuda						
Bermuda	297	104±08	12±10	88	148	58

and andesites. The Madeira archipelago represents the culmination-massif of a large submarine plateau rising 1000 m above sea level (Mitchell-Thome 1976).

Joint and river orientation measurements were made on Madeira and on Porto Santo (details in Scheidegger 2002b). The results of the statistical evaluations are shown in Table 2.10. Inasmuch as the final values of the joint maxima of the Madeira archipelago and the Canary archipelago as well as the first maxima for the stream directions on both archipelagoes are more or less identical, the two archipelagoes must be closely related: The Madeira archipelago is also affected by the anticlockwise rotation of the African plate postulated by the plate tectonic interpretation.

Azores Archipelago   The nine principal Azorean islands extend from about 36°N to 40°N and from 24°W to 32°W (Fig. 2.12). Joint and stream orientations were measured on five islands of the archipelago (S. Miguel, Terceira, Sao Jorge, Flores and Corvo), ridge trends on two of them (S. Miguel, S. Jorge); the results of the statistical evaluations are shown in Table 2.10. The archipelago straddles the mid-Atlantic ridge. Flores and Corvo are situated on the American side of the ridge, the other islands on the European/African side. The rocks (volcanics and sedimentaries) seem to be essentially subrecent to recent. The region is geologically active (earthquake of 1 January 1980 near Terceira [Geotimes 1980]).

The results of the morphometric analysis (details in Scheidegger 2002b) show that, on each island individually, at least one of stream and joint orientations corresponds to another. The trends of the ridges are normal to the main (Max. 1) trends of the streams in the two cases where such could be determined. In the whole archipelago (Table 2.10), the bulk of the joints strike essentially N-S and E-W (the common evaluations of the joints in the archipelago yield "mean" strikes of N03°E and N90°E). Presumably, the principal stresses are indicated by the bisectrices of the joint-strike maxima; viz. 47° and 137°. This fits the results from fault plane solutions of earthquakes reported by Udias et al. (1976): The mean of his P-axes referring to the Azorean area trends N146°E, comparable to the azimuth of one bisectrix of the joint-sets. On most islands, one of the stream direction maxima is directed N-S, the ridge trends E-W like the joint strikes. The orientation observations agree with the trend of most of the tectonic fracture zones (E-W and N-S) in the vicinity, and have therefore probably the same tectonic cause. The mid-Atlantic ridge does not seem to have any effect at all on the orientations of the joints, streams or ridges on the islands.

2.4.2.3
Bermudas

The Bermuda Archipelago comprises a crescent-shaped chain of about 150 islands at latitude 32°20′N and longitude 60°40′ W approximately. The entire island complex is about 30 km long and 3 km wide; it is scattered on a shallow platform rising 4500 m from the oceanic abyssal plain. This platform is, in effect, the summit of a volcanic sea mount (Stanley 1970; Stanley and Swift 1970): Beneath a layer of marine and eolian limestones around 100 m or more thick, there are volcanics consisting of basaltic lavas. Age determinations on boreholes drilled into the uppermost basalt layers have dated the emplacement of the volcanic sea mount into the middle Eocene to middle Oligocene (Gees

1969). Deeper boreholes, which presumably reached the Atlantic crustal plate, yield older ages.

The carbonate cap of the Bermuda sea mount has superficially the appearance of a coral reef. However, the "reef" has a core of eolianite indicating periods of a much lower sea level. The eolianite is reworked coral limestone; its emplacements is usually placed in the Pleistocene (Land and Mackenzie 1970).

The Bermuda island chain is oval-shaped, enclosing lagoon-type features whose origin may be volcanic as giant calderas (Gees and Medioli 1970) or giant sink-holes due to the leaching of limestone. Leaching, indeed, is visible on all the islands, as is demonstrated by the existence of natural arches or of small, round bays.

Joint orientations were measured on 19 locations on the two main islands of the archipelago; the results of the evaluation (Scheidegger, 1985c) according to the usual method, are listed in Table 2.10. The pattern is plotted in Fig. 2.12; it is seen that the stresses in Bermuda represent a simple continuation of the stresses in North America.

2.4.3
Indian Ocean

2.4.3.1
Geography/Geology

The Indian Ocean is bounded by Persia, Pakistan and India in the north, by the Arabian Peninsula and Africa in the West and by the Malay Peninsula, some Indonesian Islands and Australia in the east, and (arbitarily) by the "Southern Ocean" in the south. Its floor is characterized by various transverse and longitudinal ridges, of which the mid-Indian ridge, extending from India southward to Antarctica, is probably the most prominent. These ridges, together with fractures and trenches, appear to form a grid of "conjugate lineaments" crisscrossing the Indian Ocean floor as is, incidentally, also the case for the floor of the Pacific.

Most of the islands (not considering the evidently continental islands like Madagascar, Ceylon and Socotra) in the Indian Ocean are of volcanic origin, with the notable exception of the Seychelles: The latter may be a piece of Gondwanaland "left behind" during the drift process. Studies were made on the latter and on the Mascarenes.

2.4.3.2
Seychelles

Physical Features The Seychelles Islands constitute an archipelago between a latitude of 4° and 10°S and longitude of 46° and 57°E. Its main islands lie toward the center of a large submarine bank, the biggest ones (Fig. 2.13) are (in decreasing-size order) Mahe, Praslin and La Digue; studies were made on all three (Scheidegger 2001b).

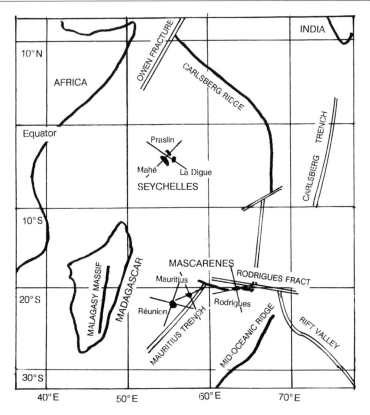

Fig. 2.13. Location of the Seychelles and the Mascarenes in the Indian Ocean between Africa and India

Geology   The islands only have a thin sedimentary cover (perhaps 500 m) upon a basement which extends downward to about 13 km depth and is commonly considered as a broken-off part of the Gondwana Shield: it consists of granite and is fringed in places by coral reefs (Baker 1963). The rocks on the *surface* of the three islands have been deeply weathered to form dark, red laterites. In many areas erosion has stripped these away so that inselberg-like rock masses were left: smooth, rounded outcrops broken by narrow, flute-like rills resembling those on limestone surfaces. There seems little doubt that these products of extensive weathering are not solely the result of contemporary erosion but that the surface morphology of the granite was defined when the sea level was appreciably lower than today (Braithwaite 1984) and was also appreciably affected by jointing (Baker 1963).

Joint orientation measurements   On *Mahe*, seven locations were investigated, mainly in granitic rocks, but joints were also observed in recent shore conglomerate and in laterite. On the island of *Praslin*, four outcrops, all in granites, were investigated on the coast as well as in the interior of the island. Finally, three outcrops, again all in granites, were studied on *La Digue*. The data were evaluated according to the standard statistical method (Sect. 1.3.3). The procedures were carried out for each island separately and also for all joints measured in the Seychelles as a whole; always for 2 and for 3 presupposed theoretical distributions. The results are given Table 2.11. As can be seen, essentially the same joint orientations are prevalent in all the Seychelles Islands, particularly if

Table 2.11. Seychelles joint strikes, river and ridge trends

Location	2 Distributions					
	No.	Max 1	Max 2	Angle	Bisectrices	
Mahe						
Joints	162	125±05	48±07	77	86	176
Links	259	143±08	72±06	71	107	17
Ridges	38	144±09	26±05	62	174	84
Praslin						
Joints	85	115±19	31±16	84	163	73
Links	141	160±08	79±06	81	119	29
Ridges	25	118±05	48±27	70	83	172
Digue						
Joints	79	130±16	52±07	78	91	181
Links	32	133±00	33±07	80	173	83
Ridges	9	168±07	5±12	17	177	87
All						
Joints	317	123±09	46±00	77	85	175
Links	691	150±00	76±00	74	113	23
Ridges	72	126±04	6±13	60	156	66

Location	3 Distributions			
	No.	Max 1	Max 2	Max 3
Mahe				
Joints	162	131±03	36±00	81±00
Links	259	143±06	26±00	86±00
Ridges	38	143±03	22±00	92±00
Praslin				
Joints	85	142±14	29±11	88±00
Links	141	136±08	2±02	78±00
Ridges	25	118±08	171±11	53±14
Digue				
Joints	79	136±12	43±07	83±12
Links	32	136±14	16±11	61±16
Ridges	9	undefined		
All				
Joints	317	136±00	37±00	84±07
Links	691	141±00	22±00	82±00
Ridges	72	127±07	169±04	37±00

three distributions are presupposed; Baker (1963) and Braithwaite (1984) have observed these same three joint sets previously during general field studies.

Physiographic considerations   Therefore we have investigated rivers and ridges. As usual, the (main) rivers/ creeks of the islands were traced from maps and divided into "links" whose trends could be measured. The resulting values were then statistically treated like those of the joints; the results are also shown in Table 2.11. The ridges were treated in a similar fashion; the results are equally shown in the table.

Discussion   An inspection of the Table 2.11 leads at once to the recognition that there is a remarkable consistency between the maximum-1 values for the joints, rivers and ridges on all islands. This consistency is particularly evident if the calculations are made for three maxima, somewhat less so if they are made for two. A reasonable consistency also exists between the maximum-3 values. This supports the earlier supposition that the joints and the physiography are closely related.

A connection with tectonics is obtained with the assumption that joints are somehow indicative of the shear of the tectonic stress field (Scheidegger 1979e, 2001a). Then the bisectrices would indicate the direction of the principal tectonic stresses: The latter would be about N-S and E-W (for two presumed joint sets). Such directions of the principal stresses could be very reasonable: An E-W stress-relief could indicate the direction of separation of the Seychelles (and India) from Africa and a N-S compression the direction of collision of Afro-India with Eurasia, in agreement with the common plate tectonic interpretations.

2.4.3.3
Mascarene Islands

General Remarks   The Mascarene islands (Fig. 2.13) form a volcanic archipelago in the southern Indian Ocean stretching from Réunion (2512 km^2; around 20°50′S, 55°30′E) along an ENE trending straight line to Mauritius (1865 km^2; around 20°10′S, 55°30′E) and onward to Rodriguez (104 km^2; around 19°40′S, 63°25′E). In spite of lying geographically more or less on a straight line, they may not be connected genetically, inasmuch as there are prominent fracture lines and deep-sea trenches separating the islands as well as separate parts of each island from one another. Details of a morphotectonic investigation of the archipelago have been reported by Hantke and Scheidegger (1998). Here, we summarize the results of this study.

Réunion   *Geologically*, Réunion, is, like all Mascarene Islands, characterized by essentially basaltic volcanism. Its formation began about 5 Ma ago; the main volcanic activity occurred about 2.1 Ma ago building up a Pleistocene shield volcano; its last lava flows occurred 13−12 ka ago. Thus, most of Réunion consists of an *old volcanic mass*. Around it, three wide basins ("cirques") have been excavated on account of the higher rainfall rate in the island's interior than at its periphery (Nunn, 1994, p. 193). The cirques drain through narrow gorges to the sea. A straight geological lineament trending N45°E separates the main (old) volcanic mass in the west from a presently active volcanic area in the east: the latter includes the Piton de la Fournaise, 2577 m high, which has been active for the last 200 ka.

Joint orientation measurements were made at 23 outcrops in all parts of the island. The data were evaluated according to the usual statistical method (see Sect. 1.3.3); the results are given in Table 2.12. *River valley* orientations were obtained by approximating their streams by straight "links" of 1 km length whose directions could be measured. These were then treated by the standard statistical method like the joints. The results of the evaluations are also given in Table 2.12. Finally, an analogous procedure was applied to *crest lines*. These "crests" were not necessarily mountain crests, but also abrupt ledges in and around the gorges and cirques. The crests seem to have three preferred directions; we list the values for the two thereof with the largest angle between them (Table 2.12).

An *inspection of the preferred orientations* shows that the joint strikes in Réunion are mainly directed N105°E and N20°E. This fits statistically the orientation pattern of the river links: The statistical evaluation gives a correspondence within 9–10°. A similar coincidence exists for the main (first) trend maximum (N128°E) of the crests with the corresponding maximum for the river links. The rivers are therefore normal or parallel to the crests. The crests are parallel to the river courses (particularly evident in the gorges around the cirques). In summary, one can say that in Réunion joints, river valleys and crests are all similarly oriented. The trend of the geological lineament does not fit the pattern of the other surface features; it appears therefore to have been created much earlier than the latter.

Table 2.12. Strike/trend directions of morphotectonic features in the Mascarene Islands

Location	No.	Max 1	Max 2	Angle	Bisectrices	
Réunion						
Joints	504	105±03	20±03	84	63	153
River links	571	115±06	29±00	86	72	162
Crests (2ex3)	222	128±00	67±00	61	98	18
Lineament	1		45			
Mauritius						
All joints	359	159±06	66±06	86	22	112
River links (2ex3)	339	158±02	34±00	83	6	96
Crests	81	160±01	64±07	85	22	112
Lineaments	2	140	70			
Rodrigues						
All joints	253	170±02	80±00	90	35	125
River links	68	161±05				
Crests	18		51±13			

Mauritius   *Geologically*, Mauritius is again wholly volcanic (Bold 1966). It contains three blocks of high country which are separated by two trench-like depressions, one trending N145°E, the other very broad depression trends N70°E. The rest of the island is plateau-like, sloping gently towards the sea. Each block of high country has a summit just over 800 m high. The island was built up during three periods of volcanism: the first in the Cretaceous/Early Tertiary (only a series of jagged stumps survive), the second in the Late Tertiary where there was a large outpouring of lava, and a third in the Pleistocene. Most of the volcanism ceased about 100 ka ago. The coastline is girt by a coral reef.

Joint orientation measurements were made at 16 volcanic outcrops. The data were evaluated according to the usual statistical method (Sect. 1.3.3); the results are shown in Table 2.12. For mophotectonic comparisons, the *River Valleys* were again approximated by straight links of 1 km length and then treated statistically. Finally, the same procedure was applied to the *crest lines* (steps again 1 km); the results are shown in Table 2.12.

Comparing the morphotectonic data, one sees that the *joint orientations* have two well-defined maxima (at 159° and 66°). The evaluations of the *river links* seem to indicate that there are more than two sets. Thus, if the statistical evaluation is made for three maxima, the main maximum of the river directions (at 158°) correlates within 2° with that of the joint strike directions. The subsidiary maxima are possibly spurious; we list that forming the biggest angle with the main maximum in Table 2.12. The two maxima (at 160° and 64°) of the *crest line directions* agree closely with those of the joint strikes. Regarding the *trends of the depressions*, one notes that one of them agrees closely (within 3°), the other vaguely (within 19°) with the joint maxima. Thus, *in summary*, one sees that there is a close connection between the orientations of joint, river, crest and lineament directions. This would seem to indicate that there is again a pronounced common (neo)tectonic control of these features.

Rodrigues   *Geologically,* Rodrigues is, like the other Mascarene Islands, wholly volcanic in origin. Its main physiographic feature is a central ridge of doleritic lava reaching 387 m in height. The island is partly surrounded by a raised coral reef studded by islets separated from the land by a shallow lagoon.

Joint orientation measurements were made at eleven locations. These were statistically evaluated according to the usual method (Sect. 1.3.3). Table 2.12 gives a summary of the results. Next, the *valley trends* were again approximated by links which were 1 km long. Because of the much smaller size of Rodrigues in comparison with the two other Mascarene islands, the total number of links is much smaller and the statistical evaluation less decisive, only one maximum is meaningful. Table 2.12 gives the values for the statistical evaluation of the data. The same procedure was applied to the *trend of the crest*, again only *one* maximum is meaningful. The numerical values are also given in Table 2.12.

A *comparison* of the Rodrigues data in Table 2.12 shows that the *river trends* agree with one of the joint maxima within 9°; there is really only one *crest line* maximum which is at an angle of 110°, i. e. vaguely at a right angles to the river directions. Thus, in *summary*, there is again a reasonable correspondence between the various geomorphic features with joint strikes, so that again a common predesign can be postulated.

Conclusions   Upon inspection of the orientation statistics of (potentially) morphotectonic features one notes that a connection exists on each individual island between the joint orientations and the orientations of prominent geomorphological features (valley courses, crest lines, lineaments). Therefore, a common morphotectonic predesign seems to be present for all these features. For the joints alone, we note that those on Réunion reflect the direction of the Malagassy massif (ca. 10°); otherwise there is a gradual clockwise rotation from Réunion to Mauritius and Rodrigues. Continuing this rotation to the location of the Rodrigues Fracture, one joint strike maximum would reach the trend of the latter. This could again be an indication of a genetic relation between the corresponding features.

2.4.4
Pacific Ocean

2.4.4.1
Morphology/Geology

The Pacific Ocean is the largest part of the Earth's hydrosphere, stretching from Australasia/New Zealand to the Americas. Its northern boundary is nearly landlocked, the only 64 km-wide Bering Strait separating Asia from North America. The southern boundary is usually put at an imaginary line along about the 50th parallel which separates the Pacific from the "Southern Ocean" encircling Antarctica. The continental boundaries of the Pacific Ocean represent an active tectonic belt with many volcanoes, earthquakes and geodetic movements.

Igneous rocks near the western border of the Pacific Ocean are mainly of andesitic type, those in the central basin of essentially basaltic (oceanic) composition; the boundary between the two regions has been called the "Andesite Line" (Marshall 1916, Chubb 1934). The studies reported here have all been made on archipelagoes of the central basin (between 20°N–20°S and 110°W–170°W, cf. Fig. 2.14), where the islands are almost entirely of basaltic volcanic origin or are due to coral growth which has been superimposed on volcanic foundations (Scheidegger 1995).

The structural pattern of the central basin involves two principal contrasting trends: (i) E-W [90°] (becoming ENE [70°] in the N, ESE [110°] in the South) fracture zones (Ma et al., 1997; cf. Fig. 2.14), these are concomitant with (W)NW-(E)SE [110°] linear ridges bearing the archipelagoes with intervening depressions (Fairbridge 1975) and (ii) N-S [0° = 180°] (NNW [160° = −20°] to NNE [+20°]) rift zones. The E-W fracture zones are commonly interpreted as transform fault systems, while the ∼ N-S rifts and ridges are assumed to constitute growth belts of volcanoes and islands (Fairbridge 1975): if the islands are taken as traces of fixed mantle hot spots, the velocity of the Pacific Plate is about 11 cm/a in a direction of WNW [110°–140°] (Guille et al. 1993). However,

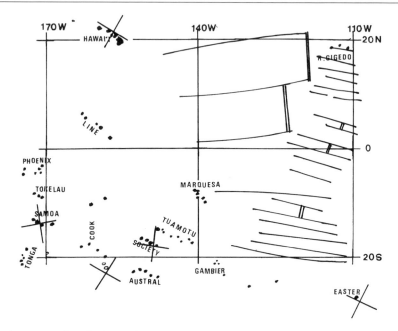

Fig. 2.14. Preferred joint strike orientations within the tectonic setting of the Pacific

hotspots may possibly not always be fixed (Pinsker 2003), which would invalidate the above estimate. The *P*-axes of earthquake fault plane solutions in the south-central Pacific [average azimuths 126° (Okal et al. 1980)] lie roughly in the direction of the ridges.

2.4.4.2
Samoa

General Remarks We begin in the western Pacific with the *Samoan Island Chain*. The latter consists of three large (from W to E: Savai'i, 'Upolu and Tutuila) and a number of small islands, for instance Manono Island located between Savai'i and 'Upolu. The three large islands are strung along a line trending about N65°W (115°). The islands W of 171°W are an independent republic (Western Samoa), those E of 171°W are U.S. territory. Studies were made in Western Samoa, in particular in 'Upolu, Manono and Savai'i (for details see Scheidegger 1995).

Morphology The *general morphological features* of the Pacific are also evident in Samoa. On *'Upolu*, the remnants of old volcanoes are visible practically everywhere, sometimes as flat-topped hills. There are lava sills that have given rise to spectacular waterfalls. *Manono Island* is a sunken volcano with coral sand beaches between promontories. On *Savai'i*, the most impressive features are the still-visible pahoehoe lava flows that occurred between 1902 and 1911 on the N-coast: the ropy structure of the solidified lava is well preserved and the damage caused by the flows is all too evident. The volcanic center from which this flow issued is still impressive. The S-coast presents a sequence of steep drops and sandy beaches; on one of the capes, a waterfall drops over a lava sill directly into the sea.

Geology   Geologically, the rocks in Samoa are principally basaltic volcanics, the oldest possibly of Pliocene age, mantled by later basalts. The youngest are from the latest eruptions which took place on Savai'i between the years 1902 and 1911 (Stearns 1975a,b).

Origin of the Islands   Various different opinions exist regarding *the origin of the islands*. According to Stearns (1975b), Western Samoa was built over a primary SE-NW rift zone and secondary E-W and N-S rifts; the latter may have been caused by lithospheric flexure (Hawkins and Natland, 1975). In contrast, Nunn (1994, p. 46) believes that the Samoan chain was originally a hot-spot trace upon which flexure may have been superimposed which has caused the recent volcanism. However, superficially, the "age aspect" in the Samoan chain appears to be reversed to the usual East-to-West ageing of the islands (cf. Hawaii, Society Islands): The youngest, presently volcanically active island of the archipelago, Savai'i, is the western-most island of the archipelago; the volcanism on 'Upolu and Tutuila in the E seems extinct and the dissection by gullies appears to increase eastward. Thus, it has been speculated (see Kear and Wood, 1959) that the Samoan island chain does not represent a hot-spot trace, but originated in consequence of an existing shear zone caused by the relative motion between the Pacific and Australasian tectonic plates.

Joint Orientation Measurements   Joint orientations were measured on three islands of the archipelago. On *'Upolu*, nine locations were investigated, all in solidified lava flows; on *Manono Island*, joints were measured in lava flows and pyroclastics at two promontories; on *Savai'i*, joints were measured at nine outcrops, one of them upon the 1911 lava flow, the rest in older solidified lavas and in pyroclastics. The data were evaluated for each island and for the entire Samoan chain according to the usual statistical method (Sect. 1.3.3); the results are shown in Table 2.13.

Interpretation for all Samoa   The joint orientation measurements on the Samoan islands show very consistent evaluation-results, although the individual errors are rather large. The results for two assumed distributions listed in Table 2.13 show: On all three islands studied, one joint set strikes between 76° and 82° (all joints: 76°), the other between 159° and 181° (1°) (all joints 170°), leading to bisectrices (i) $\sim$ 27–42° (all joints 33°) and (ii) $\sim$ 117–132° (all joints 123°). It is immediately obvious that the direction of one of the bisectrices is parallel to the trendline of the archipelago. If (ii) is taken as the tectonic compression direction, this might have induced the primary NW-SE rift zone proposed by Stearns (1975b, see above). The joint orientation data from Samoa correlate closely with those from nearby Fiji which was discussed in connection with Australasia (Sect. 2.3.3.4). Thus Samoa probably belongs tectonically to Fiji rather than to the Pacific basin.

Table 2.13. Pacific Islands: joints, lava flows – strikes/trends

Location	No.	Max 1	Max 2	Angle	Bisectrices	
Samoa						
Savai'i	173	172±20	81±17	89	36	126
Manono Is.	42	159±14	76±12	83	27	117
'Upolu	169	1±14	82±14	82	42	132
All Samoa						
Joints	384	170±10	76±10	86	33	123
Trend Archip.						115
Hawaii						
Big Island						
Joints	105	107±08	12±01	84	60	150
Lava flows		150±00	60±00	90	105	15
Maui	25	169±18	67±07	78	118	28
Oahu	105	124±01	44±00	80	84	174
Kauai	142	116±13	46±11	90	92	1
All Hawaii						
Joints	377	118±08	26±09	88	72	162
Trend Achip.		110	10			
Cook Islands						
Rarotonga						
Joints	112	120±14	35±17	84	78	168
Plate motion		115				
Society Islands						
Huahine	105	100±10	8±10	89	144	54
Moorea	63	88±12	3±12	85	46	136
Tahiti	170	102±09	12±09	90	57	147
All Soc. Isl.						
Joints	338	98±07	8±06	90	143	53
Trend Archip.		118				
Easter Island						
Joints	63	110±18	28±10	82	69	159
Lava gullies	50	68±10	152±12	84	110	20

2.4.4.3
Hawaiian Islands

General Remarks   The Hawaiian archipelago is a prominent morphological feature in the North Pacific: It consists of islands and reefs that stretch for 2500 km from about 19°N/155°W in a direction of N80°W (110°) to approx. 28°N/172°W. The chain continues underwater to the Emperor Seamount (ca. 32°N/172°E), where there is a marked bend: the feature continues as a further series of submerged seamounts ("Emperor Seamounts") extending in the direction of N10°W (170°) to about 50°N/160°E (Table 2.13). The strange "Emperor bend" in the Hawaiian complex has never really been explained: The above-water part of the Hawaiian archipelago has usually been considered as a hot-spot trace with a fixed active volcanic center being presently located beneath "Big Island" (Hawaii proper); the islands become "older" towards the NE as the plate supposedly had moved in that direction over a fixed hot spot, until they become only almost sunken shoals around Midway Island and further onward as seamounts. The strange bend near the Emperor Seamount, however, may indicate that such hot spots may not always be fixed, or else that a sudden drastic change in plate motion-direction occurred 43 Ma ago (cf. summary by Pinsker 2003); both "explanations" seem rather far-fetched. The volcanism of Hawaii is characterized by basaltic magma eruptions (cf. Poldervaart, 1971) from centers and fissures. The lava flows mostly freely, but contains some gas causing explosions that produce "ash" (Stearns, 1985) and pyroclastic layers. The constant outpouring of free-flowing lava has caused the slow buildup of high shield volcanoes.

Hawaii-Big Island   The type locality is Hawaii-Big Island, which is the newest and biggest island of the Hawaiian chain. It comprises 62% of the total land area of all the islands of the Hawaiian archipelago taken together, i. e. its area is 1.63 times the area of all the other islands combined. Its landscapes include desolate lava flows, lush coastal valleys, deserts and high mountains. It is roughly triangular in shape with sides 144 km, 120 km and 100 km long; it contains five volcanic complexes, some more than 4000 m high, about 32 km apart, connected by saddles which are 900–2100 m high, formed by overlapping lava flows. It is also supposed to be the newest island of the Hawaiian chain; one of its active volcanoes, Kilauea, alledgedly sits presently right on top of the Hawaiian hot spot (Stearns, 1985).

Joints on Hawaii/Big Island are mostly cracks in lava which have usually been assumed to be columnar cooling joints common in basaltic lava flows (DeGraff and Aydin, 1987). Their orientations have been measured at five widely scattered locations on the island and evaluated by the usual statistical method (Sect. 1.3.3). The results yield two definite preferred strike directions, viz. Max. 1 $\sim$ 107 $\pm$ 8, Max. 2 $\sim$ 12 $\pm$ 1 (Table 2.13), which are parallel and orthogonal to the Hawaiian chain and not a random pattern which would be produced by hexagonal cooling joints.

The joint orientations can be compared with the orientation statistics of the lava flows in the region of the Hawaii Volcanoes National Park (map by Decker and Decker 1986). The contours of the lava flows have been digitized into linear segments, each of 2.5 km length; making a statistical direction-analysis according to the usual statistical method (Sect. 1.3.3) yields for the trends of the segments: Max. 1 $\sim$ 250 $\pm$ 0, Max. 2 $\sim$ 60 $\pm$ 0 (Table 2.13). Again, the lava stream-edge orientations are not random, but form a systematic grid, which indicates an endogenic origin of the features (Sect. 1.2.2.2): The lava flows seem to follow pre-existing volcanic rifts and not epigenetic random channels. A comparison of the joint strikes with the lava flow directions shows that the discrepancy between the azimuths of the corresponding maxima is about half a right angle within the error limits. Crevasses also appear at an acute angle at the edge of flowing glaciers (Nye 1952), and a similar mechanism might explain in part the position of the joints with regard to the lava flows. Furthermore, if rifts arise in the principal pressure direction of the tectonic stress field, and the joints are shearing features therein, one would also have an angle of 45° between the strikes/trends of the two features (Scheidegger 1993b).

Maui Maui, the next island studied, is shaped like the figure "*eight*". It consists of two volcanoes, Haleakala, 3055 m high, and Puu Kukui, 1764 m high, connected by a low *isthmus* 10 km wide. Haleakala has a crater at its summit (last eruption: a lava flow in 1790 *a.d.*). Puu Kukui is much older (1.3–1.15 Ma old) and is marked by deep canyons radiating from its summit. The volcanism on Maui is much less active than on Hawaii/Big Island; it corresponds to a later stage of volcanism. The isthmus between the two volcanic complexes on Maui is formed of water-laid tuffs, partly cross-bedded; it has outcrops with joints. Thus, joint orientations were measured in the crater of Haleakala and on the isthmus; the result of their statistical evaluation is shown in Table 2.13. It is evident that two well-defined joint sets exist with strikes 67° ±18 and 169° ±09. The bisectrices have a direction of 28° and 118°, actually normal and parallel to the trend of the Hawaiian archipelago. This might have something to do with the motion direction of the Pacific Basin over the supposed Hawaiian Hot Spot: The tectonic compressive stress might be in the motion direction. This speculation may not be certain because it is based on solely 25 joints from two outcrops.

Oahu The island of Oahu is roughly diamond-shaped, consisting originally of two immense volcanoes which have been eroded into two rugged, parallel mountain ranges striking roughly NW-SE with a central valley in between. Oahu has supposedly been created by the Pacific Hot-Spot from which it has by now moved even somewhat farther away than Maui: The last volcanic eruption occurred on Oahu about 2 Ma ago. Since then, erosion has cut deep valleys, giving the island a "mature" look. The last post-erosional activity occurred in the Pleistocene with basalt and nephelinite flows 40–800 ka ago (Stearns 1975c, 1985).

Joint orientations were measured at five locations scattered around the island in various types of lavas. The values were then evaluated according to the usual statistical method (Sect. 1.3.3). The main result of the evaluations are two prominent conjugate joint sets with strike directions at 124° and 44°.

Kauai Kauai is the western-most island of the Hawaiian chain which is commercially accessible. Because of the distance it has moved from the supposed presently active hot spot below Big Island, its gorges and valleys have had more time (last volcanism occurred more than 3.8 Ma ago) to become spectacularly eroded: basically, they have receded forming huge scars in a region of originally coalesced shield volcanoes. Joint orientations were measured at seven locations distributed all over the island, in various types of volcanic materials, such as naturally cemented volcanic boulders, lava flows near some waterfalls at the edge of a volcanic sill, in a dry cave in volcanic tuff etc. The results from the usual statistical evaluations are shown in Table 2.13. The joint orientations correspond closely to those on Oahu.

General Interpretation of Joint Orientations in the Hawaiian Chain   The results from the four islands have been collated in Table 2.13. One notes that on *Big Island*, the joints may be mainly determined by the directions of the lava flows; however, since the latter may also be tectonically predesigned, there may be a tectonic implication in the joint orientations as well. On *Maui*, the number of joints measured was probably too small to yield a really significant result. On *Oahu and Kauai* the *joint strikes* are are (i) around 124–136°, and (ii) around 44–46°, i. e. parallel and orthogonal to the trend of the archipelago or of the supposed hot-spot trace. Similar well-defined joint maxima are obtained if a combined evaluation for the entire Hawaiian Archipelago (numerical result shown in Table 2.13) is made. This would indicate that the Hawaiian chain was formed over a shear or rift zone in the Pacific basin trending 124°–126°.

2.4.4.4
Cook Islands

Geography/Geomorphology   The Cook islands are an archipelago which extends between 8°S and 23°S and 157°W and 167°W. Its main island is Rarotonga which is a high basaltic volcanic island, roughly of the shape of an oval with diameters of about 7 (NS) and 10 (EW) km. Radioactive dating has given an age of 1.8 Ma for the last eruption. What is left of the volcano is the hard inner core (maximum altitude 653 m) and a U-shaped belt of hills sweeping from the S side of the island to the N: the latter are the remnants of a collapsed caldera. On their sides, erosion has produced fairly deep-cut valleys and spectacular waterfalls. The island is fringed by a reef with about 6 passages through it leading to the lagoon on the sea side of the coast. On the land side, there are alluvial plains which have been cultivated by Polynesians with tropical crops for hundreds of years.

Geology   Geologically, Rarotonga is usually considered as part of the trace of a mantle hot spot. It lies in the prolongation of the belt of the Austral Archipelago (in French Polynesia). An active center of the latter is MacDonald Sea Mount at its SE extremity, but, inasmuch as the velocity of the Pacific Tectonic Plate is supposed to be about 11 cm/a in a direction N65°W (from N115°E) and inasmuch as the distance between MacDonald Sea Mount and Rarotonga is 1000 km, the latter is about 10 Ma too young to have been caused by the MacDonald active center. To save the hot spot hypothesis, one has postulated, in view of several identified volcanic buildup periods on other islands of the same belt at the W end of the Australs (12.5 Ma and 0.6–1.9 Ma ago), several hot spots or an entire "hot line" on the same belt (cf. Guille et al., 1993, p.18).

Joint Orientation Measurements   Joint orientations were measured at six locations across the island in basaltic lava rock. The evaluation was carried out according to the usual statistical procedure (Sect. 1.3.3). Table 2.13 shows the results.

Interpretation   Regarding a possible interpretation, one can say that one of the joint sets seems nearly parallel, the other orthogonal to the direction of the Pacific Ridges and also to the supposed motion direction (115°) of the Pacific Plate. This would actually contradict any idea that Rarotonga has anything to do with a volcanic hot spot trace, but rather relate to one of the shear lines of the Pacific ocean basin.

2.4.4.5
Society Islands

General Remarks   The Society Islands are part of French Polynesia which consists of several archipelagoes. The three southern ones (from N to S: the Society Islands, the Gambiers with an extension to Pitcairn and the Australs with extension to Rarotonga) form more or less linear archipelagoes which are roughly parallel amongst each other striking about NW-SE (N115°E) and are separated by deep oceanic trenches. Generally, within one and the same group, high islands featuring peaks consisting of basaltic volcanic rocks are situated in the SE, low atolls in the NW and mixed islands in the middle. They have been interpreted as traces of mantle hot spots; their K-Ar ages range from 0 to 10–12 Ma, increasing from SE to NW corresponding to a motion of the inert Pacific tectonic plate over fixed hot spots in a NW direction: The hot spots supposedly caused rift zones on which the islands of the archipelagoes are now located (cf. Guille et al. 1993).

The *Society Islands* in particular are considered as a hot spot trace. The ages of the basaltic volcanic rocks increase systematically in a roughly NW direction from the present center at the island of Mehetia (Guille et al. 1993). From the values of the ages and the trend of the archipelago one calculates an approximate velocity of the Pacific Plate of 11 cm/a in a direction of about N65°W (i. e. *from* N115°E) in this region.

Tahiti   Tahiti is the largest island in the Society Islands. Its shape resembles a figure eight lying on its side, inasmuch as it is built of a double cone of olivine basalts, connected by a low isthmus and surrounded by a single barrier reef with a few passes enclosing a lagoon. The larger cone, Tahiti-Nui, lies to the NW, the smaller, Tahiti-Iti to the SE. Both cones preserve their general initial shape, but are deeply gullied, the precipitous gorges having been graded to Pleistocene low sea levels. Around the coast there are "fossil cliffs" in places, up to 300 m high with waterfalls and caves, fronted by a narrow Holocene coastal plain based on raised reefs and littoral sands covered by alluvial deposits: These have been planted to tropical crops or developed into inhabited places (e. g. the city of Papeete) and parks. The river mouths are generally swampy. In Tahiti-Nui the volcanic core contains a plutonic plug of nepheline monzonite, theralite and syenite which has been recognized as clearly an oceanic basaltic differentiate. Otherwise, the lavas are dominantly alkaline basalts; they were highly fluid and for the most part deficient in pyroclastics (Chevalier, 1975a). Joint orientations have been measured at eight locations around and in the interior of the island.

Moorea   Moorea is the next island of the archipelago just NW of Tahiti. It is of a triangular form and is, in effect, part of a huge caldera, of which only the southern part (the apex of the triangle, pointing towards Tahiti) has remained; it is ringed by a reef with few passes to the lagoon fringing the island. The north side (the "base" of the triangle) is open to the interior of the caldera; thus there is a semi circle of mountains, the rim of the caldera, which becomes more and more impressive as one approaches the island from the sea. The highest mountain (Mt. Tohiea, 1207 m) represents the well-preserved former volcanic plug; the island is tilted over to the SW. The old lavas, basalts, phonolites and minor pyroclastics are extensively affected by laterization (Chevalier, 1975b). Joint orientations were measured at three locations.

Huahine   Huahine, the next island to the NW, is a double volcanic island (Huahine Nui in the N, Huahine Iti in the S), divided by a narrow, shallow (an isthmus at low tide) strait, across which a bridge has been built, but united by single rectangular fringing reef with dimensions of ca. 15 km × 10 km. Around the two volcanic plugs, the largest elevation reaching 680 m, land masses extend like low, swampy snake-like arms enclosing spectacular bays. The dominant rock is labradorite basalt, with minor trachytes and phonolites. The lavas are deeply weathered to laterite and kaolinite. Shore terrasses consist of raised coral (Chevalier 1975b). Joint orientations were measured at five locations.

Interpretation   Joint orientations were measured on the three islands as indicated. These were then evaluated according to the usual statistical method (Sect. 1.3.3); the results are shown in Table 2.13 for the individual islands as well as for the archipelago as a whole. It is at once seen that the azimuths of the joint strikes on the three islands vary from 88° to 102° for the first and from 3° to 12° for the second maximum. If Moorea, on which only 3 outcrops were investigated, is left out, the variation is even much smaller: from 100° to 102° resp. from 8° to 12°. If "All Society Islands" are taken as *one* location, the joint strike maxima are (i) 98 ± 07 and (ii) 8 ± 06. The direction of the strikes joint set (i) is vaguely parallel to the trend of the island chain and thus also parallel to the supposed motion direction (115°) of the Pacific plate, the other orthogonal thereto. Guille et al. (1993) have identified the direction of N115°E also as a seismic rift zone.

2.4.4.6
Easter Island

Geography   Easter Island (Spanish Isla de Pascua, Maori Rapa Nui) lies, far from any other land, at around 28°10′S latitude and 109°20′W longitude. Its area is about 130 km^2; its shape is that of a rectangular triangle with its apex in the north and its hypotenuse of 22 km running ENE. The height of the triangle is 12 km.

Geology   The island is volcanic in origin and owes its triangular shape to apparently extinct volcanic centers that form its three corners. The highest lies in the north (Maunga Terevaka 506.5 m), the next, Maunga Puakatiki (370 m), on the east and the third, Rano Kau (324 m), in the west. Maunga Puakatiki was formerly a separate island surrounded by cliffs. The most southerly of the volcanoes (Rano Kau) was the main source of the material used to make the celebrated stone statues. The island is principally covered by volcanic ash, but there are also many lava flows, particulary on the southern side of Maunga Terevaka as well as many parasitic tuff cones in radical lines, one containing obsidian. There is no record of volcanic activity within historic times; the latest eruption has possibly occurred around 400 *a.d.* The volcanic rocks are essentially oceanic basaltic (the island lies well on the W side of the Andesite line), they are relatively siliceous and nepheline-free; they include oligoclase-andesite. There are no streams, but a series of steep-walled lava gulleys on the flanks of Maunga Terevaka form prominent geomorphic features. The shores of the island are cliffed, but no terracing is evident. There are no coral reefs; the winter sea-water temperature is too low (Richards 1975).

Joint orientations   A total of 63 joint orientations were measured at three outcrops. The joints were evaluated according to the usual statistical method (Sect. 1.3.3). The results are shown in Table 2.13. These were compared to the directional patterns of the lava gullies: the latter were digitized into 500 m-links and again statistically evaluated; the results are also shown in Table 2.13.

Interpretation   It is interesting to note that the situation is exactly the same in Rapa Nui as on Hawaii/Big Island: the lava gullies are in the directions of the bisectrices of the joint maxima. The explanation therefor is presumably the same as that for the situation on Hawaii (cf. Sect. 2.4.4.3). The joint strikes are, on the other hand, parallel to the usually supposed motion direction (from 110°) in the Pacific basin. This is particularly remarkable inasmuch as Easter Island is commonly not assumed to lie on the Pacific plate, but on the Nazca plate to the E of the former.

2.4.4.7
Regional interpretation of observations

A regional interpretation of the results of all observations in the Pacific Basin must relate the latter to the *structural pattern* of that basin which is characterized (cf. Sect. 2.4.4.1) by two principal directional trends: (i) E-W (70°–110°E) directed fracture zones, commonly interpreted as transform fault systems, which are concomitant with linear ridges bearing the archipelagoes and intervening depressions; and (ii) N-S (−20° [160°] to +20°) rift zones. The islands are usually considered as traces of fixed mantle hot spots, the velocity of the Pacific Plate being about 11 cm/a in a direction of WNW [from 110°–140°]. The *P*-axes of earthquake fault plane solutions in the south-central Pacific [average azimuths 126°] lie roughly in the direction of the ridges.

As far as the *Samoan Islands* are concerned, it is noted that the age pattern is reversed from that of the other islands (age increasing E to W instead of W to E) and that the joint orientations do not mark the trend of the island chain: The latter agree closely with those from nearby Fiji (cf. Tables 2.7 and 2.13). Thus Samoa probably belongs tectonically to Fiji rather than to the Pacific basin.

In all other investigated Pacific archipelagoes (*Hawaii, Society Islands, Raro-tonga and Easter Island*), the azimuths of the preferred joint strikes vary from 98° to 120° for the first maximum, and from 8° to 35° for the second (Table 2.13). The direction of the first joint set is roughly parallel to the direction "(i)" in the Pacific Basin, i.e. directions of the (transform) faults and island-traces (and also to the [supposed] plate-motion direction [110°–140°]), the other orthogonal thereto, i.e.parallel to the direction "(ii)" above, which corresponds to the rift zones assumed that constitute growth belts of volcanoes and islands. Thus, it appears that the linear island-traces might have been formed over shear/rift zones in the Pacific basin trending ca. 120° and may not correspond to a hot-spot trace of a fixed mantle hot spot. The situation is further aggravated by the sharp "Emperor bend" in the Hawaiian complex. This might indicate that something might really be wrong with the idea of fixed hot spots altogether (cf. Sect. 1.6.3; Pinsker 2003).

2.5
Global Morphotectonic Conclusions

After collecting and presenting global morphotectonic data from all over the world, it is possible to consider their implications. So far, we have shown that there is a general correspondence between the orientation of large scale geomorphic features of a region (river courses, ridges) with the orientations of joints. Regardless of any theory of the genesis of either rivers, ledges or joints, this correspondence makes it likely that there is a general relation between these features. Inasmuch as there is a general agreement that joints have been created somehow by tectonic stresses, one can infer that the other features mentioned have also been created by tectonic processes.

These relations can be extended to the whole Earth. Many morphological features form patterns on the globe. The lineaments in the eastern Pacific present almost an E-W/N-S grid (cf. Fig. 2.14). Buser (1966) has analysed morphological features in Africa which he considered as "paleostructures", such as the Nigerian anticlinorium which strikes ESE-WSW. Generally, many features seem to have a propensity to strike or trend more or less E-W and N-S: This is certainly the impression one obtains when inspecting the figures in Part II of the book. Storetvedt (2003) hypothesizes that *all* fracture/joint systems on Earth are caused by its rotation and therefore trend predominantly N-S/E-W; he gives many qualitative examples.

Inasmuch as there is much variation in the joint orientations in detail, a suspected pattern can only be determined by statistical means. We have therefore taken the results of the data evaluation for the various regions discussed in this book and analysed them further statistically. Thus, we have subjected the orientation maxima for the joint strikes of the individual regions of a whole continent in turn to the Kohlbeck–Scheidegger (Sect. 1.3.3) analysis. The results of this procedure are shown in Table 2.14. Then, the continents themselves have been grouped into Laurasia and Gondwanaland; we have proceeded similarly also with the oceanic island data. Finally, the regions of the world have

Table 2.14. Strike directions of joints for World Regions

Location	Regions	Max 1	Max 2	Angle	Bisectrices	
Laurasia						
Europe	22	169±12	84±14	85	126	36
Laurent.Asia	12	13±11	105±29	88	149	59
North America	19	166±09	78±19	87	122	32
Arctic regions	6	153±19	63±22	90	18	108
All Laurasia	59	170±06	82±11	87	126	36
Gondwanaland						
Africa	5	166±16	82±20	84	124	34
Penins. India	11	1±04	92±02	88	137	47
Australasia	6	8±16	97±17	90	142	52
Northern S. Am.	8	5±14	101±21	83	142	53
Lesser Antill.	11	161±23	72±12	89	117	27
Southern S. Am.	4	132±22	43±23	88	178	89
Antarctic Isl.	4	5±15	93±12	88	139	40
Antarctic Pen.	3	24±00	123±08	81	163	73
All Gondwana	52	4±03	95±07	88	139	50
Oceans						
Atlantic	5	12±17	102±18	90	147	57
Indian Ocean	6	147±16	64±22	83	105	16
Pacific Ocean	5	18±19	105±18	87	152	62
All Oceans	16	16±14	102±12	86	59	149
Whole world	127	2±03	94±00	88	138	48

Fig. 2.15. Direction-rose of the joint strikes of all world regions considered in this book. The directions are taken as parameters (degrees N>E) in the ordinary geographic coordinate system regardless of the latitude- and longitude-position of the corresponding region

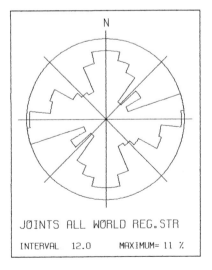

been processed as one group. The result of this procedure is interesting: The larger the base area for the statistical analysis, the closer are the orientation maxima of the joint strikes to N-S/E-W, and the smaller become the individual fluctations and errors (Table 2.14 and Fig. 2.15). This result is particularly remarkable because N-S, say in Spitsbergen (10°E) – is *not* parallel to N-S e. g. in Greenland (50°W) – these N-S directions differ by roughly 60°! Thus, a word of caution has to be added: The results of Table 2.14 depend very much on the coordinate system used on the Earth for listing orientations. If the poles of the coordinate system would be chosen e.g. at Singapore (90°E at the equator) and its antipole in the Pacific Ocean instead of at the present rotation poles, the values in Table 2.14 would be quite different: the directions of fiord trends/joint strikes in Baffin Land and Greenland would not be so similar, but differ quite a bit. Nevertheless, inasmuch as such good correspondences result if the usual North-South geographic coordinate system is used, there is an indication that the rotation of the Earth may play a role in the genesis of the joints.

The above conclusions are based on factual data. They are independent of any views regarding the specific (shear or tension) origin of the joints, and independent of any geotectonic hypotheses. It is not the aim of the present book to discuss how our results may affect the acceptability of the latter.

CHAPTER 3

Local Morphotectonics

3.1
Scope of Chapter

After having looked at the continent-wide aspects of morphotectonics, it is now our task to turn to its local aspects. The latter refer to river valleys, basins, coast lines, isolated hills, volcanic features and mass movements.

Unfortunately, it is often difficult to separate the purely local from the general global significance of particular morphological features, so that we have occasionally already discussed some local morphological forms if they happened to have implications with regard to global tectonics. We shall, however, try to keep overlapping dissertations to a minimum and refer to the earlier chapters when necessary.

3.2
Local Valley Morphology

3.2.1
Introduction

The visible indented channels on the surface of the Earth in which water, debris or ice (and possibly: lavas) flow, are called "valleys". In principle, these flows occur under the action of gravity, but "traffic jams" behind or tidal waves can also cause an additional push or even a "push" in the opposite direction.

The customary view of the generation of valleys is that a river is eroding the sole and the sides of its channel, cutting downward and sideward, thereby forming a valley. This has been in contrast to a view held widely in the 18th century that regarded the valleys as tectonic features, i. e. gigantic chasms (Fränzle 1968). The erosional origin of river valleys is to date a practically universally accepted theory. Nevertheless, there is plenty of evidence that *purely* erosional activity cannot be the whole story. Rivers *do* follow tectonic structures, so that one can speak of a tectonic predesign (or co-design) of many morphological patterns found in the landscapes shaped by rivers. It is our aim (without denigrating the considerable effect of river erosion) to point out the (neo!)tectonic influences in present-day river morphology.

3.2.2
Drainage Patterns

River patterns in drainage basins correspond to topological networks (cf. Sect. 1.5.3. In particular, in most cases (if there are no bifurcations) they can be represented as a topological tree (arborescence); its root is at the outflow point from the basin.

In relatively featureless sedimentary areas of low relief the formation of junctions in the drainaige network is a random process; Horton's (1945) laws of stream numbers can be derived theoretically by calculating the probability of the occurrence of corresponding graphs (Liao and Scheidegger 1968) in a random set of graphs. The theoretical results agree well with observed conditions in nature, such as with the Wabash drainage system in the flat American Midwest (Ranalli and Scheidegger 1968).

The topology of a typical drainage network on a single, uniform slope into a main valley can also be explained by statistical reasoning, viz. by a random walk model of fluvial erosion and capture which can be generated on a computer (Scheidegger 1967b). Such random patterns agree closely with observations in sedimentary areas of low relief.

In areas of medium to high relief the drainage patterns have often been considered as antecedent to the orogeny. Thus, Staub (1934; see also Ahnert, 1996, p. 254) postulated that the drainage pattern of the rivers in the central plain of Switzerland is the carbon copy of an antecedent purely "erosional" Miocene drainage pattern in a plane. However, one notices that the individual links in drainage patterns are often parallel to the joint-strikes in the area: We have given many examples of the correspondence of joint strikes and river directions from all over the world in Chap. 2 of this treatise. For the details, we refer the reader to the tables provided there, particularly those pertaining to mountainous areas, such as the Alps or the Himalayas. Thus, it must be observed that the drainage orientation pattern in areas that are not completely plane and featureless is to a large extent determined by neotectonic processes; naturally, erosion widens and deepens the tectonically "predesigned" valleys.

3.2.3
River Courses in Plan

Rivers hardly ever flow along a straight line, rather, their courses follow wiggly lines (cf. Sect. 1.5.2). In flat, *unstructured plains*, the rivers form large loops (meanders) which swing back and forth, generally never remaining in any one place for very long periods of time, unless interfered with by man. In *bedrock plains*, one speaks of incised meanders inasmuch as the loops become more stationary within deeper gullies. In *mountainous country*, the rivers run along sinuous courses with rock spurs alternating from the two sides.

The causes of the sinuosity of river courses have been sought in the general hydraulics of rivers, in the dynamics of channel flow, in the stochastic presence of obstacles and in morphotectonic effects.

The hydraulics of rivers presupposes that there is a well-determined relation between an equilibrium river-bed slope and other hydraulic variables
(bed load, flow velocity etc.). Generally, a river running along a straight line
would have a bed slope which would be too steep for equilibrium conditions:
Thus the river has to form a longer course which would necessarily be meandering. The dynamics of channel flow could lead to meandering inasmuch
as a completely channel-parallel flow would become unstable. Thus helicoidal
cross currents develop leading to lateral bank erosion causing the river to form
the meanders. A different approach to the problem of meander formation is
based on statistical considerations: The meander train is assumed to be the
result of the stochastic fluctuations of the direction of flow due to the random
presence of direction-changing obstacles in the river path. One can then try
to calculate the most likely path (Langbein and Leopold, 1966) or may attempt
to find the expected path (random walk; this problem has been investigated
by means of a physical model [Thakur and Scheidegger 1970]). In both cases,
typical meander trains have been obtained. The stochastic theory of meander
formation works very well and has been tested in flat, basically unstructured
areas (Thakur and Scheidegger 1970) such as the plains of Illinois (Salt Fork
Vermilion River) or Manitoba (Assiniboine River).

The above theories all assume a completely unstructured territory. However, such territories are rather rare; in most cases, some tectonic precondition
(evidenced by the presence of joints) is present. This is quite obvious in mountainous areas, and probably also in meanders "incised" (stabilized in deep
gorges) in flatter areas, such as the Aare loop around Berne, or the "Goose-
necks" of the San Juan River in Utah, U.S.A. Such incised meanders have
commonly been assumed to have been caused (Holmes 1944, p. 197; Holmes
1965) by the rejuvenation of a landscape containing ordinary plains-meanders
in only a thin cover of easily eroded deposits; the deepening channel is etched
into the underlying (solid) rocks with the original winding course being preserved. Ahnert (1996, p. 220) suggests that the development of such meanders
is aided by the presence of vertical structures in the underlying rocks; Hantke
(1991, p. 170) postulates that the loop around Berne is determined by shear
faults.

A case study to demonstrate the tectonic origin of incised mountain river-
meanders has been made with regard to the course of the Tamina River which
enters the Rhine near Bad Ragaz, Canton of St. Gall in Switzerland (Hantke and
Scheidegger 2001), after running for about 6 km in zigzags through a gorge
(sketch map in Fig. 3.1). The river segments obtained by walking dividers
of 125 m length along it were digitized and their orientation structure was
determined. Two of the three resulting orientation maxima coincided within
3 degrees with the two strike maxima of the joints measured at nine outcrops
in the area (cf. Table 3.1). This points towards a common origin of joints
and river directions and therewith to a predominantly tectonic origin of the
meanders.

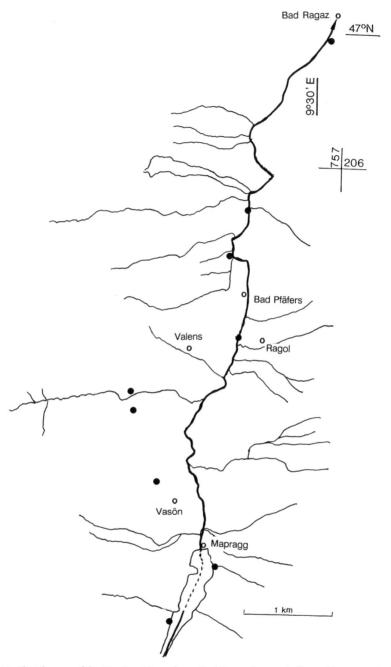

Fig. 3.1. Sketch map of the Tamina River above Bad Ragaz (towns indicated by *open circles*); the nine outcrops where joint orientations were measured are marked by *black dots*. River segments of 125 m length were taken along the main river marked by a *heavy line*; tributaries are indicated by *thin lines* (After Hantke and Scheidegger 2001)

Table 3.1. Tamina Gorge, Strike (resp. Trend) Directions

Feature	#	Max. 1	Max. 2	Max. 3
River segm. 3 distrib.	66	132 ± 03	39 ± 00	180 ± 03
Joints 2 distrib.	188	129 ± 06	38 ± 16	

3.2.4
Problems of Longitudinal River Profiles

It is well known that tectonic events have a significant influence on the *longitudinal profile of a river*. Thus, the cataracts of the Nile have obviously been caused by the intrusion of igneous materials (granite) into the surrounding country rock (Nubian Sandstone; Said, 1962); and the Falls of the Rhine near Schaffhausen by a tectonically predesigned river channel intersecting a ledge consisting of banked Jurassic limestone above an old river bed (Hantke, 1993 p. 121).

These qualitative statements can be quantified by a numerical comparison of actual river profiles with theoretical ones that would be the result of purely "exogenic" processes. Geomorphologists have assumed that an equilibrium river has a longitudinal profile that is exponential; however, the theoretically smooth profiles have in reality often been found interrupted by the presence of "knickpoints" whose initiation is generally ascribed to some geomorphic instability. The classical view is that equilibrium is not possible around a knickpoint: the latter must therefore wander upstream and become obliterated (cf. the presentation of the corresponding argument and literature by Scheidegger, 1991, p. 209). Thus, stable knickpoints cannot be caused by the "exogenic" river action alone.

In this context, a detailed numerical analysis of the river profile of the longitudinal profile of the Rhine river in Europe has been made. Figure 3.2 shows a plot of this profile (original data from Ahnert, 1996); the kink at km-250 represents the Falls near Schaffhausen, the long horizontal stretch above it the Bodensee. We have then calculated and drawn (Fig. 3.2) the exponential curve which best fits the data; it is obvious that no exponential curve can be made to fit the actual data: The head water regions are over-steepened which can only be explained by the continuing uplift of the Alps; moreover the knickpoint at km-250 *does* exist and does not wander upstream.

A further confirmation of these contentions has been obtained by Maiti (1980, p. 60; 1991, p. 114 ff.; see also Mukhopadhyay, 1982) who analysed river profiles in West Bengal at the base of the foothills of the Eastern Himalayas: he showed that no theoretical geomorphic profiles can be made to fit the actual data; knickpoints correspond to the traces of upthrust planes; moreover, the

Fig. 3.2. Longitudinal profiles (actual profile: *heavy
line*; and best-fitting,
theoretical-exponential profile: *thin line*) of the Rhine
River, West-Central Europe.
Ordinate: elevation (km)
above sea level; abscissa: distance (km) of thalweg from
source. Note knickpoint at
km-250 (Falls of the Rhine
near Schaffhausen) and
oversteepened head water
region in the actual profile

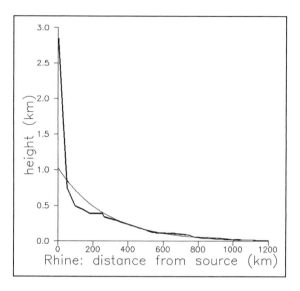

head water regions are over-steepened which can only be explained by the
continuing uplift of the Himalayas.

In addition, Pizzuto (1992) has shown that quite generally the size grading
of the bedload particles in natural rivers cannot be produced by sorting and
abrasion; the sizes must be supplied by the downstream tributaries which again
points to tectonic influences.

In summary, many valley features are, contrary to the "established" opinion,
not solely caused by exogenic agents, but are to a large extent tectonically co-
designed.

3.2.5
Transverse Valley Profiles

3.2.5.1
Introduction

The transverse cross-sections of fluvial valleys have generally been ascribed to
purely erosional and intrinsic fluvial processes. Thus, it is usually maintained
that the "natural" or "original" (youth) form of a valley is V-shaped (Davis
1924, p. 90; Holmes 1944, p. 153; Holmes 1965, p. 479; Ahnert 1996, p. 277). As
their evolution progresses, valleys become widened and deepened a fact which
is ascribed to the scouring and steepening of the channel sides by the river
itself, to rainwash and gullying, and to soil creep (Holmes 1944, p. 160 ff.).
Particulary glaciers are thought to be apt to transform V- into U-valleys.

3.2.5.2
Scouring Power

However, now there is overwhelming evidence that the erosive action is insufficient for the creation of the valley profiles (Scheidegger and Hantke 1994): It is a simple hydromechanical fact that the erosive power of a river is much too small to cut itself into solid rock (Hantke 1991, p. 297); indeed, the hydraulic bed shear in a river is at most of the order of 100 kPa (Magilligan 1992), whereas the shearing strength of rocks lies in the order of 10 MPa (cf. e.g. Scheidegger 1982a, p. 188–189 for a summary), i.e. at least two orders of magnitude higher. On phenomenological grounds, Hantke (1978 p. 284; 1991 p. 264) has noted that the rivers in the folded Jura mountains follow mainly geological lineaments and Barsch (1969 p. 209) indicated that the erosive power of the rivers in that same area is, in spite of a considerable significance of periglacial processes, much smaller than commonly thought, without giving any specific quantitative rates, though. Heuberger (1975) found that the downward incision into solid rock caused by the outflow from a lake formed some 8000 yrs ago by the Koefels (Tyrol) slide was only a few meters since the slide occurred, i.e. less than 0.5 mm/a – much too little in order to keep up with the Alpine uplift rate of several mm/a. Furthermore, in dated Australian lava flows (in the Tumbarumba district, N.S.W.) Young and McDougall (1993) have shown that most major streams have incised by only 5–18 m/Ma; similarly, in field studies in the gorges of Kaua'i (Hawai'ian Islands) Seidl et al. (1994) have found erosion rates of 10–100 m/Ma. Such rates are far too small to have much geomorphological significance in so-called "youthful" areas where the uplift rates are of the order of several km/Ma. Thus, a river can *never* cut itself into solid rock – except under very special circumstances: if cavitation is involved (possibly in a water fall) or if the erosion does not occur in the solid rock at all but in a soft layer underlying it.

Thus, river genesis must be influenced by the tectonic stress-induced faulting and jointing and the underlying tectonic structure generally: river valleys follow shear faults, split-up anticlines, heads of layers, fronts of nappes and break through mountain chains along pre-existing tectonic faults (Hantke 1978, 1991, 1993 p. 8).

3.2.5.3
Glacial U- versus Fluvial V-Valleys

The story that glacial valleys are supposed to be U-shaped, fluvial valleys V-shaped, goes back at least to Davis (1909, 1924). Since that time it has been quoted in practically every textbook on geomorphology or on physical geology (eg. Machatschek, 1952, p. 128–129; Holmes, 1944, p. 204–252). Undoubtedly, the erosion by flowing ice has had a pronounced effect. Some ice-scoured valleys in Scandinavia are indeed U-shaped; however, there are many V-shaped glacial valleys in Switzerland (Hantke, 1991, p. 44): an U-shape may only be *simulated* by the debris in the valley floor, since the valley sides in flat-lying carbonate rocks break off almost vertically and the debris accumulates at the valley floor. In such cases, the original valley shape is evidently tectonically predesigned (Gerber 1945): The cause of the morphology are "tectonic lineaments", i. e. faults and joints (see also Sonder 1938). Thus, the erosive action of flowing ice has generally been overestimated, and it is not tenable to assume that all V-shaped valleys are of fluvial, all U-shaped valleys of glacial origin (Hantke and Scheidegger, 1999).

3.2.5.4
Overdeepening by Fluvial/Glacial Erosion

Glaciers, or possibly glacial rivers, are not only supposed to have caused U-shaped valleys, but also the *overdeepening of valleys*, including *piedmont lakes* (Penck and Brueckner 1909); this, too, has become the generally accepted view (e. g. Holmes, 1944, p. 224; Ahnert 1996, p. 337). Actually, a glacio-*fluvial* origin of these valleys can be excluded, because the fill in them is of glacial origin (Schluechter 1987). Actually, the assumption that glacial or fluvial erosive action causes the excavation of valleys, was the cause of the disaster on 24 July 1908 during the building of the Loetschberg-tunnel in the Bernese Oberland: masses of water-logged gravel were encountered unexpectedly 160 m beneath the present valley floor; the resulting flood caused the death of an entire shift of workers in the tunnel (Schluechter 1983; Labhart 1991, p. 126). Evidently, the existence of loose gravel at such a depth was (and is) unconceivable in the light of the theory that a glacier or a river had excavated the valley. In fact, the primary depressions and clefts owed their existence not to erosion, but rather to various tectonic processes. The same holds true for the "overdeepening" of piedmont lakes on the N-side of the Alps in Europe (Hantke, 1991, p. 117) as well as on the E-side of the Andes in Peru (Dumont, 1993). Thus, it is not the glacial action alone, but rather a fundamental tectonic design which is responsible for the genesis of such features.

3.2.6
The Origin of Gorges

3.2.6.1
Introduction

Gorges are deeply incised, often spectacular features on the Earth's surface, usually with a river at their bottom. Their genesis has generally been ascribed solely to the erosive action of the river itself; but, like in all river valleys, the erosional power is insufficient for the deepening of valleys in solid rock (cf. Sect. 3.2.5.2). It is the purpose of this section to show that the river only erodes pre-existing crags; tectonic stress-induced faulting and jointing must be the primary reason for the genesis of gorges.

3.2.6.2
Morphology of Gorges

Phenomenologically, there are three types of gorges which can be distinguished, viz. transverse gorges, watershed gorges, and plateau-edge gorges. They can be defined as follows:

Transverse gorges are gorges that cut transversally through a mountain or hill range that appears as a barrier across a river course, such as the Aare Gorge in the Bernese Oberland or the "Cluses" in the Jura Mountains of Switzerland. The standard explanation of such gorges is that an antecedent river retained its course during the genesis of the mountain range concerned by "sawing" its way into the rising rock barrier (see e.g. Cotton 1968; or any textbook on geomorphology, e.g. Louis 1979, p. 340).

Watershed gorges are usually claimed to have been formed by rivers draining from rising slopes during the genesis of a mountain range, continuously regressing so as to form gorges back to the precipitous valley heads. Classical examples of this type of gorge are the Minster Gorge, in the Swiss Alpine foreland hills or the Samaria Gorge in Crete. Incidentally, the concurrence of two such gorges at the crest on opposite sides of a watershed by erosive regression had also been proposed as a possible genesis mechanism of transverse gorges (Louis 1979, p. 343), but is no longer seriously considered.

Finally, *plateau-edge gorges* are thought to have been formed at the edge of a plateau owing to lithological conditions: if competent rocks overlie soft materials, a waterfall may be the result which regresses back into the plateau, leaving a gorge in its wake. The plateau-edge may be caused by a fault or may be a simple erosion-scarp (cuesta-crest) in mildly dipping strata. A famous case of this type is represented by the Niagara Gorge at the border of Ontario (Canada) and New York (U.S.A.), others by some "cluses" in the Swiss Tabular Jura mountains.

3.2.6.3
Origin of Gorges

Standard explanation   Summarizing, the standard explanation for the genesis of gorges is in all cases that the latter are the result of the rivers maintaining original, antecedent courses during the evolution and change of the terrain by "sawing" themselves into the substratum.

Morphotectonic explanation   As described in Sect. 3.2.5, transverse river profiles and gorges resulting solely from erosive processes are untenable. As an explanation of the genesis of gorges it is therefore proposed that they are neotectonically designed, corresponding to tectonic processes that are active to the present day (Scheidegger and Hantke 1994). This theory is supported by a statistical analysis of the orientation structures of the river segments at the bottom of the gorges and comparing the former with the orientation structure of the faults and of the steeply-dipping, i.e. non-lithologic joints in their vicinity. It is generally seen that the orientation patterns of the river segments and that of the joints in the area agree with each other quite well, whence it is inferred that the gorges are basically predesigned and caused by neotectonic processes: the rivers simply follow pre-existing cracks and zones of weakness.

3.2.6.4
Paradigms of the morphotectonic origin of gorges

a. Transverse gorges
Aare Gorge   The Aare Gorge, in the Canton of Berne in Switzerland, connects the valley widenings of Innertkirchen and Meiringen across a transverse barrier (the Kirchet-Riegel) formed of Upper Jurassic (Malm) to Cretaceous sandy–calcareous sedimentary rocks (location map in Fig. 3.3). The standard explanation for the genesis of this gorge, as of all such gorges, is that an antecedent river cut itself into the rising barrier maintaining its previously existing course (Müller 1938; Labhart 1991). However, a morphometric analysis of the Aare gorge showed that there is a significant tectonic influence in its formation. First, Hantke (1991) identified older buried alternative river gorges across the Kirchet Riegel which certainly could not have been antecedentally cut into the barrier at the same time as the present river channel. Similarly, the walls of the gorge are not vertical but oblique, which is quite contrary to any theory that they had been cut vertically by the river. Next, Hantke and Scheidegger (1993) made a detailed statistical comparison of the directions of joints and river segments in the region of the Aare Gorge (rose diagrams of joint strikes and river segments in Fig. 3.4) which showed that these directions coincide. Thus, the Aare river did not cut itself into the growing Kirchet-Riegel, maintaining its "original" course, but followed (variously at different times) pre-existing joint clefts, mainly on the level that connects the valley widenings of Innertkirchen and Meiringen. In this instance, only the joint clefts were cleaned out: the erosion at the foot of the gorge is minimal. It is even possible

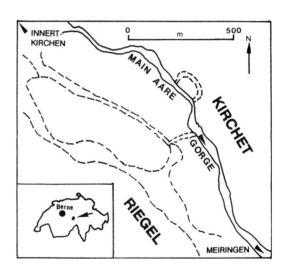

Fig. 3.3. Map of the Aare Gorge (modified after Hantke and Scheidegger 1993)

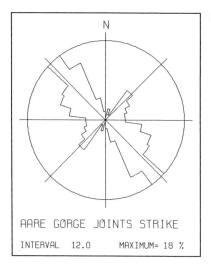 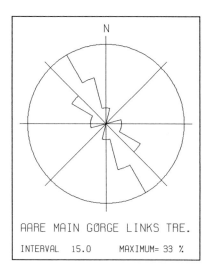

Fig. 3.4. Aare Gorge area: Joint strike rose on *left*, trend rose of the 50 m-segments of the main gorge on *right*

that the pre-existing joint clefts reached quite deep down and were not only not "cut" by the Aare river, but were filled up with sediment to the present "base level of erosion".

Cluses and Related Forms in the Swiss Jura Mountains   A second group of paradigms of transverse gorges is represented by the cluses in the Jura Mountains of Switzerland, where generally SW-NE striking longitudinal structures determine the landscape pattern: The anticlines form mountain chains, the synclines longitudinal valleys and basins. Every now and then, gorge-like features, the so-called *cluses*, break through the mountain chains. Again, their genesis has commonly been ascribed (cf. e.g. Labhart 1991) to antecedent rivers maintaining their courses during the rising of the anticlines. This view, however, is untenable in the light of the general remarks on the origin of gorges. Indeed, it has again been observed that the orientation structure of the segments of the rivers flowing in the cluses is non-random, which is (Sect. 1.6.1) indicative of a non-exogenic origin of them. In addition, the orientation structure of the cluse-rivers correlates with that of the joints in the area. Regarding the fact that the joints are known to have been caused by recent tectonic processes, the same must be assumed for the cluses: the latter owe their genesis to complicated geologic lineaments, folds and shear faults (Hantke and Scheidegger 1994).

b. Paradigms of Watershed Gorges
Minster Gorge   A typical watershed gorge is represented by the Minster valley between Oberiberg and Gurgen in the Canton of Schwyz in Switzerland where it forms an impressive limestone gorge (map in Fig. 3.5). Geologically, the area

Fig. 3.5. Minster Valley between Oberiberg and Gurgen in the Canton of Schwyz in Switzerland: towns indicated by *large black dots*, locations of joint orientation measurements by *small black dots*. Rivers shown by *heavy lines*, roads by *thin lines*. Study area is situated between the locations marked "BEGIN"and "END"

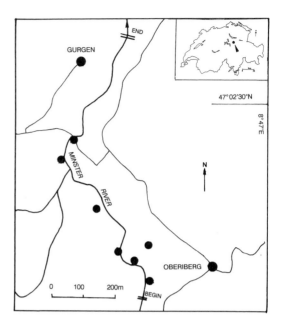

is dominated by rocks of one of the Helvetian nappes (the Druesberg-nappe), which marks the edge of the Alps by steeply ascending layers and folds. Large masses of ultrahelvetic and subalpine flysch are found in the depressions of this nappe; occasionally single penninic masses ("klippes") are superimposed on the latter. Finally, the whole region has been affected by glacial action during the recent ice ages. Flysch overlies the limestone which caused the landscape to have a slide morphology. The Minster River flows in zigzags which presumably follow pre-existing joints. In order to test this assumption, joint orientations were compared with the orientations of the 25 m-river segments in the area, using the usual statistical method (Sect. 1.3); Fig. 3.6 shows the resulting joint strike and river trend (25 m segments) roses. A non-parametric inspection of the latter shows that two main maxima (N-S and E-W) are similar for river segments and joint strikes. Thus, there cannot be much doubt that the gorge and the joints are genetically connected. Furthermore, the joint orientations correspond to the "European" directions, implying that the genesis of the joints and of the gorge was conditioned by the Central European tectonic stress field.

Samaria Gorge   Another example of a watershed gorge has been discussed in detail by Scheidegger and Hantke (1994): the Samaria Gorge in the western part of the Greek island of Crete. The latter is part of the South Aegean arc connecting the Hellenides of the Greek mainland with the Taurides of SW Anatolia. This arc was thrown up as a mountain range during the Tertiary Alpine orogeny (Jacobshagen 1986). In Crete, on its S side, some spectacular watershed gorges have evolved, of which the Samaria Gorge is the most easily accessible. It appears to zigzag in its course, indicating from a visual standpoint

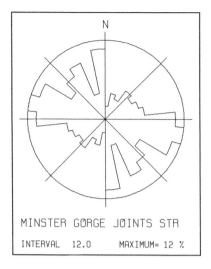
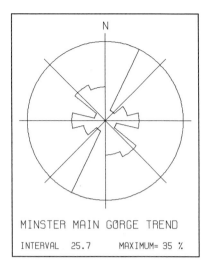

Fig. 3.6. Minster Gorge: Joint strike rose on *left* and the river trend rose (25 m – segments) on *right*

a tectonic (following conjugate joint systems) rather than a purely erosional origin (Scheidegger and Hantke 1994). Joint orientations in the gorge were compared with river segment orientations and it turned out that the main river trend ($65° \pm 00°$) is equal to one maximum ($63° \pm 21°$) for the joint strikes, which indicates that the course of the Samaria River gorge has probably been designed by the same neotectonic stress field that caused the joints. Furthermore, the joint orientations around the Samaria gorge are the same as in the rest of Crete (Sect. 2.2.1); thus the Samaria gorge has evidently been caused by *global* tectonic processes (Scheidegger and Hantke 1994).

c. Paradigms of Plateau-Edge Gorges

Niagara Gorge   The gorge below the Niagara Falls, at the Canadian-US border between lakes Erie and Ontario, is *the* classical example of a plateau-edge gorge. According to the commonly accepted view, a resistant capping formation (massive middle Silurian dolostone) outcropping along an escarpment (Niagara escarpment) is underlain by soft lower Silurian limestones. Rapid erosion at the base undermines the cap and the falls retreat upstream along their original course maintaining a vertical or undercut face, leaving behind downsteam a canyon and kettle-like whirlpools eroded by eddies. The rate of retreat is presently about 1 m/year (Holmes 1965 p. 524, Tovell 1979, Tinkler 1993). This is the common view. However, even a very cursory examination of a map of the area (Fig. 3.7), indicates prominent kinks in the trend of the Niagara Gorge. Furthermore, a buried gorge (St. David Gorge) continuing in the direction indicated by the "Whirlpool" has been found to the W of the present gorge. This makes it doubtful that the "Whirlpool" has been cut by

Fig. 3.7. Map of the Niagara Gorge area. Loca-
tions of joint orientation
measurements shown
as *black dots* [modified
after Tinkler (1993) and
Scheidegger and Hantke
(1994)]

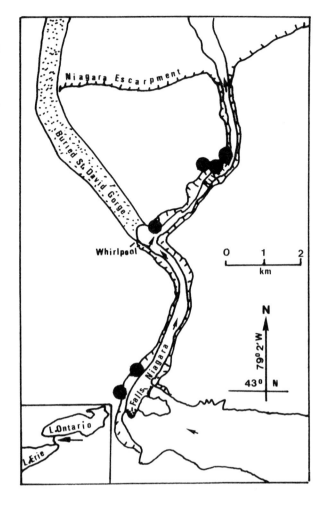

eddies as it raises the question why and how the Niagara River has cut *two*
gorges "antecedentally" into the plateau above the Niagara escarpment. Thus,
physiography would seem to suggest that the river and the gorge(s) follow pre-
existing fracture lines. This impression has been confirmed (Scheidegger and
Hantke 1994) by statistical evaluations of the joint orientations measured at the
six outcrops indicated by black points in Fig. 3.7, and comparing them with the
statistics of the gorge trends obtained by measuring the trends of the 500 m-
segments of the present and buried gorges. Figure 3.8 shows the rose diagrams
for the two features; one notices immediately that there is a good agreement
between the (*non-parametric*) main direction-maximum of the joints and that
of the gorge(s) (both 150°). This orientation, incidentally, also corresponds
more or less to one of the regional joint orientation maxima "North of Lake
Erie" (164° ± 00° calculated parametrically; Sect. 2.2.3); the corresponding Ni-

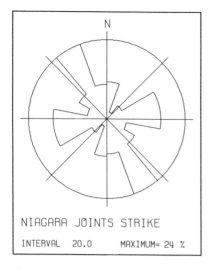
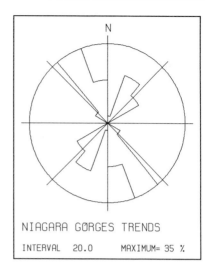

Fig. 3.8. Niagara Gorge: Strike rose of the joints *on left* and trend rose of the 500 m-segments of the river *on right*

agara joint maximum calculated *parametrically* would actually be 158° ± 09°, which is very close indeed to the regional one. Therefore, one would again conclude that the course of the (buried and presently visible) Niagara gorges has not been determined by erosive action by an antecedent river, but has been primarily determined by the tectonic stress field which also caused the joints to form.

There are also plateau-edge gorges *in the Swiss Jura Mountains.* We may take as a paradigm the Gorges du Seyon between Valangin and Neuchâtel (Hantke and Scheidegger 1994): Joint orientations were compared with valley trends in these gorges. A comparison of the corresponding strike/trend roses made the coincidence of the main direction maxima immediately obvious. Thus, one would again infer an identical genetic origin of the joints and of the gorge: Both have been predesigned by neotectonics.

3.2.6.5
Discussion and Conclusions

We may summarize the essential statements of this section as follows (Scheidegger and Hantke 1994):

First, except under very special circumstances (occurrence of cavitation), the erosive power of a river is too small by at least a factor of 100 for the water to be able to attack solid, tectonically undisturbed rock.

Second, the orientation structure of the river segments in gorges is generally *not* random, but shows strongly preferred directions. This fact by itself falsifies the thesis that gorges have been exogenically eroded. Exogenic processes would have produced random features (cf. "Principle of Antagonism", Sect. 1.2.2.2).

Third, the orientation structure of the gorge (river) segments conforms generally to that of the joints in the vicinity of the latter. This strongly suggests (cf. Sect. 1.6.1) a neotectonic origin of the gorges.

3.3
Basins

3.3.1
General Remarks

Basins are relatively depressed regions on the Earth's surface with a more or less equidimensional outline (Holmes 1944, p. 414 ff.). The term is applied to all kinds of sags of the crust. Ideally, the drainage in a basin is inwards, but in many basins drainage exits through gaps in their rim, so that this simple criterion cannot be applied generally. The term "basin" is also given to ancient crustal sags which have been filled by sediments and possibly uplifted into plateaus. Basins may occur just in the dry landscape, or they may contain bodies of water in their interior. For us, the most interesting basins are those of tectonic-structural origin: in addition to the control by ancient structures, their present-day morphology has been shaped by recent neotectonic processes. In the following sections, we shall show some paradigms of landscape and lake basins.

3.3.2
Landscape Depressions

3.3.2.1
Introduction

First of all, we shall discuss basins that are simply depressions in the landscape, Generally, their origin is structural: old faults, grabens, edges of nappes etc. Nevertheless, they, too, have been affected by recent tectonic processes, a fact which expresses itself mainly in a discordance between the tributary river directions and the orientations of such basins. Our paradigm here will be the Vienna Basin.

3.3.2.2
Vienna Basin

Structural background   The Eastern Alpine ranges disappear eastward under the Tertiary and Quaternary sediments of the ("Inner-Alpine") *Vienna Basin* (Boegel and Schmidt 1976) which represents a deep basin that trends obliquely to the Alpine zones. The lowering of its bottom occurred at the beginning of the Middle Miocene (Thenius 1974, p. 76), it reaches its greatest depth (about 5.5 km) 18 km to the east of its rim.

Fig. 3.9. Sketch map of the western edge of the Vienna Basin (*dotted*) with outcrops visited (*black dots*) and river courses investigated (*thick lines*). Modified after Scheidegger 1998b

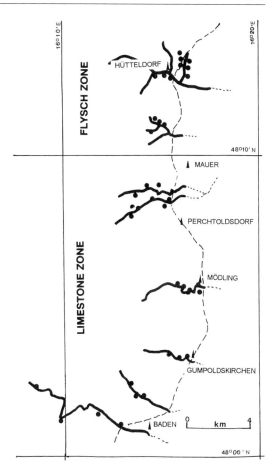

Recent morphology   The steep slopes on the western rim of the Vienna Basin are cut by several rivers and creeks in gulleys and gorges, of which seven (with their tributaries) are most prominent (cf. Fig. 3.9). The question arises as to whether the latter have been caused by pure erosion (cutting into the rock) of the water, or whether the creeks were following primarily neotectonic preconditions.

Exogenic interpretation   According to the generally accepted view (cf., e. g., Ploechinger and Prey 1974), these gulleys and gorges have been formed as erosion channels which follow the shortest route eastward from the Alps, cutting the SW-NE trending edge of the Alpine Vienna Woods in an acute angle, resulting in transverse valleys.

Neotectonic interpretation   However, it is well known that small creeks can never "saw" their way through solid rock so as to form a gorge (cf. Sect. 3.2.5.2): Their erosive power is much to small. Thus, the creeks at the edge of the Vienna Basin with their gulleys and gorges cannot have been formed by "pure erosion", but must have been formed by following neotectonic faults and clefts. To support this view, a comparison of the orientations of joints and valleys was made (Scheidegger 1998b).

Joint orientation measurements In total, 842 joint orientations were measured at 38 outcrops in the seven prominent valleys shown (with their tributaries) in Fig. 3.9. The bedrocks north of 48°10'N in the investigated region were flysch-sandstones and schists of the middle and upper Cretaceous (Brix 1970 p. 101; Brix 1972; Ploechinger and Prey 1974), those in the south of 48°10'N were limestones and dolomites dating from the Triassic (main dolomite) to the Paleocene (Ploechinger and Prey 1974). The data were treated statistically accordling to the method explained in Sect. 1.3. Parametrically, preferred directions of N102° ± 00°E and N13° ± 00°E are found (see also Table 3.2). This, incidentally, is also quite close to the strike directions (cf. Table 3.2) for the joints in Eastern Austria (Scheidegger 1979a).

Valley trends In order to compare joint strikes with valley orientations, the river stretches marked with black lines in Fig. 3.9 were digitized in 96 segments of 500 m length; the directions of the latter were treated statistically in the usual manner. Parametrically, one obtains as preferred directions N102° ± 06°E and N48° ± 18°E (cf. also Table 3.2).

Discussion A comparison of the joint strikes with the valley trends shows that the first direction maximum of the joint strikes (104°) agrees closely with that of the river trends (102°). The second trend maximum of the valleys can hardly been considered as defined; the parametric evaluation gives a doubtful result, in view of the large error as well as in view of the fact that the two maxima are only 54° apart; they are not "conjugate" and there is, in fact, only *one* prevailing river trend, at an acute angle to the rim of the Basin. Inasmuch as the first (main) direction trends of valleys and joints are only 2° apart, it can be stated that this supports the hypothesis that the river directions, like the joints, are determined by *neotectonic* (post-Miocene) processes, and not by "pure erosion". This agrees also with the time of the creation of the Vienna basin which is supposed to have occurred in the Upper Miocene.

Table 3.2. Strike/trend directions of joints and creeks in the Vienna Basin

Feature	No.	Max. 1	Max. 2	Angle	Bisectrices	
Joints						
Vienna Basin	842	104 ± 00	13 ± 00	89	148	58
E-Austria	16reg	116 ± 22	27 ± 17	88	161	71
River segments						
Vienna Basin	96	102 ± 06	48 ± 18	54	75	165

3.3.3
Lake Basins

3.3.3.1
Introduction

Some basins on the surface of the Earth are filled with water. In this case, one is speaking of "Lake Basins". We have a similar situation as in the case of other terrestrial basins: Their origin is generally structural, but the present-day morphology has been very much affected by neotectonics. We present some examples.

3.3.3.2
Alpine Lakes

The Alpine lakes of central Switzerland are partly surrounded by high rock walls, partly embedded in softer molasse landscapes. At any rate, they are like a collection of pearls in four particular regions, the first containing Lakes Aegeri, Lauerz and Zoug in a *molasse landscape*, the second containing the various arms of the *Lake of Lucerne* in Helvetian nappe rocks, the third containing a string of lakes in the *Obwalden Valley*, and the fourth containing three partially artifical *lakes at the NE-edge of Central Switzerland*. Figure 3.10 shows a sketch map of the region in question.

Regarding the origin of these lakes, we note that Heim (1893) considered the respective basins as drowned old river valleys which subsided due to the crustal warping caused by the load presented by the nascent Alpine mountain ranges.

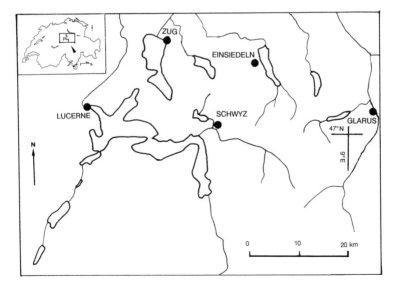

Fig. 3.10. Sketch map of the Alpine lakes (*heavy outlines*) of Central Switzerland

Alternatively, glacial erosion was widely considered as the cause of these lake basins (e. g. Kopp 1962). Finally, Hantke (1991 p. 182) proposed that the nucleation of them lay in tectonic processes occurring during the Alpine orogenesis, such as splitting nappe-edges and nappe-fronts (particularly regarding the Lake of Lucerne).

To address this problem, Hantke and Scheidegger (2003) made a field study of the morphotectonics of these Alpine lakes, including of their drainage areas. Particular attention was paid to the trends of the lakes and the evident transsection of the shore hills by the tributary creeks: The latter flow seldom in the direction of the direct, steepest descent. A comparison of the directions of these three features showed the following facts:

1. There is a general correlation between the strike-directions of the joints and the trend-directions of the tributary creeks/rivers in central Switzerland; the method for the comparison was statistical as outlined in Sect. 1.3. Numerically, one obtains directions maxima for the 4609 joint strikes measured of N163°E and N80°E, for the 1273 river segments measured N155°E and N67°E (cf. Table 3.3). The errors of the values come out as quite low because of the large number of data, so that the agreement must be considered as close. It is therefore very likely that there is a genetic connection between the origin of joints and river valleys.

2. There is no general correlation between the trends of the lake basins (parametric: N116° ± 20°E and N23° ± 28°E (cf. Table 3.3); thus, lake basins and joints must have been formed largely independently of each other: since it is known that the joints are of very recent (post-Miocene) origin, the nucleation of the lake basins must have taken place before the occurrence of the change of the tectonic stress regime (as postulated e. g. by Laubscher 1987) in the Miocene. Thus, the lake trends must have been predesigned by the late Miocene emplacement of the Alpine nappes. The basins were enlarged subsequently by the scouring action of the glaciers during the ice ages.

3. Since there is a correlation between the joint strikes and the river directions, the latter do not correlate with the lake directions, either. The rivers/creeks, like the joints, have been delineated by neotectonics.

Table 3.3. Strike/trend directions of joints, creeks and lakes in Central Switzerland

Feature	No.	Max. 1	Max. 2	Angle	Bisectrices	
Joints	4609	163 ± 00	80 ± 00	83	122	32
River Segments	1273	155 ± 00	67 ± 00	88	111	21
Lake Trends	18	116 ± 20	23 ± 28	87	160	70

3.3.3.3
Great Lakes (N-America)

As a further paradigm, we may mention the studies by Eyles and Scheidegger (1995) that have been by made around the Great Lakes of North America, notably in the sedimentary areas north of Lake Ontario and north of Lake Erie (cf. Sect. 2.2.3.3): The postglacial rivers trend mainly parallel to one of the joint sets (N-S) and normal to the other (Table 2.3 and Fig. 2.4). Similarly, even the preglacial bedrock channels, covered by thick deposits (up to 200 m) of glacial and interglacial sediments, also run parallel to the joint strikes (Table 2.3). According to the arguments presented in Sect. 1.6, this suggests that the *dominant* influence on the trend of buried bedrock channels in the western Lake Ontario basin is the recent geotectonic stress field and associated regional joint system, independently of the genesis of the lake itself. Given that modern river valleys are cut into a thick cover of Pleistocene sediment but show strikingly similar trends, it can be suggested that the drift cover is extensively fractured and that such fractures have controlled the development of the modern drainage system. One concludes that bedrock joints, buried bedrock channels cut across Paleozoic strata, and postglacial rivers cut into Pleistocene drift in south-central Ontario, have been determined by the mid-continent neotectonic stress field possibly in existence since the Late Jurassic. Bedrock channels and modern river courses reflect a common tectonic control and not any simple "epigenetic" erosional history involving an "antecedent" drainage pattern.

3.4
Shore/Coast Lines

3.4.1
General Remarks

The next local geomorphological forms to be considered are shore- and coast lines. In this instance, mainly exogenic processes, such as wave- and ice-action have been advocated as determinant factors in their genesis. We will attempt to demonstrate, by presenting specific examples, that neotectonic forces have also a most important influence.

3.4.2
Scarborough Bluffs

Morphology   The first example is one concerning a lake shore: the Scarborough Bluffs on the North shore of Lake Ontario, Canada (Eyles and Scheidegger 1999). Just to the East of the City of Toronto, the shore of Lake Ontario is formed for 16 km by a remarkably straight line of bluffs whose highest points reach (Eyles et al. 1985b) some 85 m (i.e. altitude 155 m above mean sea level) above the modern lake level (altitude 75 m above mean sea level). Behind the

Fig. 3.11. Sketch map of the location of the Scarboro Bluffs (modified after Eyles and Scheidegger, 1999)

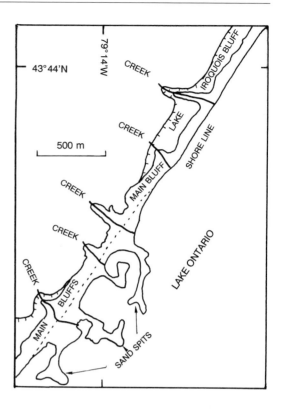

Bluffs, a plateau stretches to the North. The Scarborough Bluffs proper generally terminate at a terrace about half-way up to the plateau above. A second set of bluffs, often only visible as much degraded, somewhat steep slopes, are seen above this terrace: These have been caused by an earlier (much bigger) lake (Lake Iroquois) which came into existence about 12,000 years B.P. (Hough, 1968): Approximately 14,000 years ago, the land on which Scarborough is situated, was covered by a 2 to 3 kilometers thick sheet of ice. As the ice slowly melted, it formed the Great Lakes as we know them to-day, but there were several intermediate stages where the lakes were quite different in size and shape. Each lake, particularly Lake Iroquois, left an imprint on the landscape which is visible today: wave action created shorelines quite distinctive in shape and easily identifiable; thus Lake Iroquois caused the lower terrace of the Scarborough bluffs. Average rates of retreat of the Bluffs landward lie presently between 0.31 and 0.76 m/a (Eyles et al. 1985b). Figure 3.11 gives a sketch map of the area.

Geology (see Eyles et al., 1985a; also Eyles and Eyles, 1983)   The Lake Ontario basin is filled by sediments from the penultimate (Illinoisian/Riss) glaciation and last interglacial (Sangamon) stage. These old sediments are blanketed by a thick sequence of sediments from the last glaciation (Wisconsin/Würm) which is exposed in the Scarborough Bluffs: The glacial and interglacial sequences of the Toronto area fill a broad bedrock basin, trending NS, that connects Lake Ontario to Georgian Bay. This complex basin has been identified by Spencer (1890) as the precursor of the modern St. Lawrence River, which, as the "Laurentian River", flowed directly across the Ontario Peninsula from Georgean Bay to Lake Ontario. The area is underlain by middle Ordovician bedrock (Rogojina, 1993) whose nearest exposures are found in the valleys of the Rouge and Little Rouge rivers (cf. Fig. 3.11).

Stratigraphy   The exposures along the Scarborough Bluffs represent, in effect, a window into the infill of the Laurentian Channel. Thus, the succession along the bluffs is composed essentially of a lowermost delta body (Scarborough formation) draped by a glacial complex of diamicts (pebbly muds) and intervening deltaic lithofacies. A lower prodelta member of the Scarborough Formation delta, about 30 m thick, is composed of laminated silts and clays with many graded fine-sand units. The delta top is some 35 m above the modern lake level at Scarborough. Several abandoned channels up to 100 m deep and 1 km broad are cut into the delta-top and are infilled by a fine-grained diamict that has a drape-like geometry. The latter is overlain by sandy deltaic lithofacies which also includes two more diamict units.

Origin of the Bluffs   The *basic* genesis of the Scarborough Bluffs originates from a level-rise of Lake Ontario by about 23 cm/century. This causes recession of the shore line and hence bluffs. The lake-level is rising because of differential glacial rebound: the ice at the outlet of Lake Ontario was thicker than at the inlet, hence the rebound at the outlet (St. Lawrence) is faster causing a "damming up" of the lake and thus a rise of the lake-level. Thus, the general thesis is that the cliff-recession is caused by stress-relief owing to the ongoing post-glacial uplift. However, the general decay pattern (albeit NOT the actual erosion which is certainly of exogenic [hydraulic] origin) may also have been influenced and co-designed by the neotectonic intra-plate stress field.

Procedure   This possibility was tested by a comparison of the joint orientations measured in the Pleistocene material of the Bluffs with those in the bedrock at the lake-bottom. Indeed, prominent subaqueous joints are seen as black lines – filled cracks – when peering from a boat to the lake bottom (glacial clays) when peering downward; the orientation of their strikes could easily be visually determined.

Results and conclusions   The data were evaluated by the usual statistical method (Sect. 1.3); then a comparison was made not only of the strikes of the joints in the bluffs with those on the lake-bottom with each other, but also, in addition, with those in the bedrock of the Rouge Valley, with the trend of the bluffs and the trend of the creeks. Table 3.4 shows the results. One notices that bluff-, lake-bottom- and bedrock-joints have more or less the same orientations, as far as they could be defined. Furthermore, the trends of the creeks in the area and of the shoreline also fit into the same scheme. This would indicate an identical origin of bedrock-, lake- and bluff-joints, which can only be the present-day local tectonic stress field: The "scaling" decay of the bluffs

Table 3.4. Scarborough Bluffs: Strike/trend directions of morphological elements

Loc.	No. of Meas.	Max. 1	Max. 2	Bisectrices	
Joints					
Bluffs all	42	134 ± 06	45 ± 00	90	180
Lake Bottom	21	132 ± 07			
Bedrock Rouge	602	118 ± 02	31 ± 04	71	161
Trends					
Creeks	5	126 ± 08			
Shoreline	1		40 ± 00		

and therewith the orientation of the bluffs and shoreline is NOT only induced by stress relief and subsequent hydrological effects, but is also controlled by tectonics.

3.4.3
Fiords

As already noted in Sect. 2.2.4, the genesis of fiords (Holtedahl 1967) is generally thought to be due to "glacial overdeepening". However, in the discussion of the morphotectonics of Spitsbergen (Sect. 2.2.4.2), it was shown that there are indications that fiords are of tectonic origin and that the glacial influence was negligible during their genesis: Orientation data of joints were compared with the trend of the fiords, and it was observed that the joint strikes are close to the fiord trends; their rectilinear patterns correspond to the conjugate shear lines of the neotectonic stress field (details see Sect. 2.2.4.2).

Incidentally, similar observations have been made with regard to fiords in Greenland and Baffin Island: In both locations, joint strikes and fiord trends are parallel to each other (cf. Table 2.4), this points to a common, i.e. tectonic origin. In addition to arguments presented in Sect. 2.2.4, one may now consider "glacial overdeepening" to be mechanically impossible (cf. Sect. 3.2.5.4).

Thus, in all probability, fiords are primarily of tectonic, and not of exogenic, origin. The glaciers have only deepened and excavated preexisting tectonic features.

3.5
Inselbergs

3.5.1
General Remarks

A rather common occurrence is the presence of hills that stand apparently rather alone and are detached from their neighbourhood. Such isolated hillocks and peaks are particularly common in arid landscapes; Holmes (1944) has imported the German term "Inselberg" (island mount) into English to name such features. In (semi-)deserts, such *"classical" inselbergs* have particular properties, such as an onion-like scaling on their surface.

However, isolated hillocks are not only characteristic of desert regions, but occur also in other types of landscapes, although they lack there the particular desert characteristics. Various names (such as "monadnocks" and others) have been used to designate isolated hills in non-desert climates, but we shall stick to "inselbergs" as generic term for all such objects. Non-desert inselbergs are particularly frequent in *piedmont areas*, where hill ranges disappear into the plains. Often, "pieces" of such ranges become detached and become isolated hummocks.

A further, rather particular form of inselbergs are *escarpment outliers*: hills paralleling an escarpment; they are evidently (geologically, morphologically) part of the escarpment, but have somehow become split off from the latter.

Finally, *glaciers* are believed to have produced many types of isolated hummocks, such as roches moutonnées and drumlins.

Regarding the genesis of inselbergs, denudation and erosion (chemical, aeolian, glacial, fluvial) have generally been assumed as principal causes (cf. e.g. Ahnert, 1996). However, the contention of the author is that the unusual forms of most of these solitary mounts have been co-designed by tectonics; a support of this conjecture is obtained by a comparison of the directions of the joints on these features with the directions of other geomorphological elements, such as creeks, gullies and hill trends in their vicinity. The general result of such comparisons is that the joint strikes correlate with at least one of the morphological directions in the same area: Inasmuch as the tectonic origin of joints is beyond question, this confirms that the morphology has been co-designed by tectonics. We shall present some typical examples of the categories mentioned above.

3.5.2
Classical Inselbergs

3.5.2.1
General Remarks

Geography   As noted, the term "inselberg" has first and foremost been invented as a designation for the isolated hills that are frequently found in arid zones, particularly of Africa. We take the northern part of Nigeria as a type area (Scheidegger and Ajakaiye 1985), roughly the region from the latitude of Zaria ($11°05'N$) northward to its border with Niger (map in inset of Fig. 3.12). This region is a dry area at the southern edge of the Sahara.

Geology   The entire area is underlain by a migmatitic basement, 2–3 billion years old, which had variously been intruded by "older granite" batholiths some 650 Ma ago (Panafrican intrusions, cf. Sect. 2.3.1.5). This "older granite" has been eroded in the form of "inselbergs", where there is the controversy whether these are the results of differential erosion of an earlier solid surface,

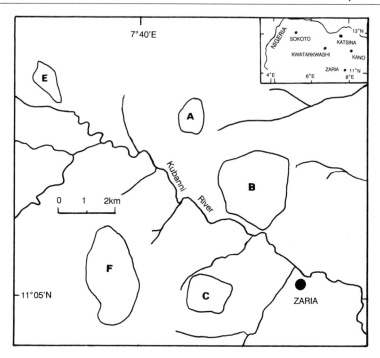

Fig. 3.12. Map of northern Nigeria showing (*inset*) location of the Kwatarkwashi Inselberg and (*main figure*) the area around Zaria with five inselbergs. All bergs except *E* consist of "older granite", *E* of biotite gneiss

or whether they are remnants of rocks that have been intruded in this form at depth. At the location where the granite does not come to the surface it is covered by a laterite crust which may have been transported some distance as it contains granite pebbles. The meteorization occurs according to the Instability Principle: Holes, once started, tend to become bigger. Then, sheeting and breakup along the joint system is induced leading to bizarre forms which are called "Tors". On gulley sides, slip scars ("blaiken") may be observed.

Morphology   The morphology of the bergs is classical: They present themselves either as "whalebacks" or properly rounded inselbergs. Sometimes there are several levels of granite domes superposed upon each other.

Tectonics   The inselbergs have been affected by tectonic events. Thus, one finds evidence of shear zones; the gulleys are evidently tectonically controlled as they zigzag along fracture systems. The tectonic events have been ancient (as evidenced by displaced dykes) or recent (evidenced by slickensides on recent joint surfaces).

3.5.2.2
Kwatarkwashi Inselberg

Kwatarkwashi Inselberg is located about 170 km to the NW of Zaria on the road to Sokoto, somewhat N of the latitude of Kano (Fig. 3.12, inset) in northern Nigeria. It is one of the biggest inselbergs in the country. It consists of "older" granite and forms part of a granite complex containing also other inselbergs. It rises very abruptly and very steeply above the village of Kwatarkwashi. The ledge to the summit is rather steep and shows much exfoliation. The flanks of the berg show evidence of rock falls. The meteorization produces tor(statue)-like forms. Jointing is prominent, leaching often being initiated by the joints.

The joint orientations measured at two outcrops of the berg line up with the general directions found in northern Nigeria (see Table 3.5; cf. also Sect. 2.3.1.5). It is therefore permissible to suggest that the joints have been determined by the global tectonic stress field in the central part of Africa.

Table 3.5. Joint strikes on inselbergs and river trends in N Nigeria

Loc.	No.	Max. 1	Max. 2	Angle	Bisectrices	
Kwatarkwashi inselberg						
Joints	46	155 ± 08	56 ± 15	81	15	105
Zaria region						
Inselberg joints						
A	24	94 ± 24	15 ± 16	78	54	144
B	21	116 ± 01	1 ± 13	65	149	58
C	20	96 ± 24	0 ± 18	84	138	48
E	24	107 ± 06	35 ± 34	72	71	161
F	23	87 ± 20	5 ± 15	82	46	136
All	112	100 ± 06	7 ± 08	87	143	53
Rivers						
500 m links	96	138 ± 13	63 ± 06	75	100	10
North Nigeria						
Joints	4reg	164 ± 20	83 ± 20	81	123	33

3.5.2.3
Zaria Inselbergs

A series of prominent inselbergs surround the city of Zaria in northern Nigeria. They are indicated and marked by the letters A-C and E-F in Fig. 3.12. Inselberg E consists of biotite gneiss, the others of "older" granite (Wright and McCurry 1970).

The orientations of joints have been measured at all five inselbergs and statistically evaluated according to the standard method (Sect. 1.3.3), for the individual bergs as well as for the group of them. In addition, the trends of the rivers in the region (marked by heavy lines in Fig. 3.12) have also been statistically analyzed (links of 500 m length). The results are shown in Table 3.5. It is observed that the individual inselbergs line up with regard to the joints, but they do not fit together with the "normal" direction for north Nigeria (Table 2.5). The rivers do not fit together with the joints either, but they do fit (within the rather large error limits) with the general northern Nigerian joint strike directions.

Inasmuch as there is a correspondence of the joint orientations on all five inselbergs, it appears that they have a common origin: Here there must be a *local* anomaly of the tectonic stress field that caused them. Inasmuch as the river directions correspond to the *regional* conditions, it is likely that there is a difference in age between the tectonic stresses that caused the joints and those that affected the river valleys.

3.5.3
Piedmont Inselbergs

3.5.3.1
General Remarks

As noted, piedmont inselbergs are usually broken-off parts of chains of hillocks at the ends of mountain ranges that disappear into the flat areas adjacent to the latter. They occur at the foot of all great mountain ranges; here we give examples from Austria and Switzerland.

3.5.3.2
The Vienna (Austria) Region

Setting   The eastern Alps disappear in the vicinity of Vienna eastward under the Tertiary and Quaternary sediments of the Vienna Basin (cf. Sect. 3.3.2.2). Affected by faults and folds, the ranges broke off into some remarkable lonely mounts, such as, amongst many others, the hill upon which the monastery of Göttweig stands. In five individual objects, correspondences have been found for various morphological elements (Scheidegger 2002a), viz. at the Kaltbründlberg in the Lainzer Tiergarten, the Bisamberg just beyond the city limits, the monastery hill of Göttweig, and finally the Hundsheimer Berg and the Braunsberg near Hainburg (cf. map in Fig. 3.13). The inselbergs are located in various geological regions: in the Flysch zone of the Vienna Woods, in the cystalline Moldanubian gneisses, in granites which intruded the latter, and finally in extensions of the Little Carpathian Mountains, in which scattered sediments surround a crystalline nucleus (Thenius 1974).

Procedure   As noted, the conjecture is that the pronounced isolation of the mentioned "inselbergs" is primarily caused by the neotectonics of the region. This conjecture can be tested by comparing the directions of joints, streams and other morphological elements. This is done in the usual way by making a statistical analysis of the features in question as described in Sect. 1.3.3.

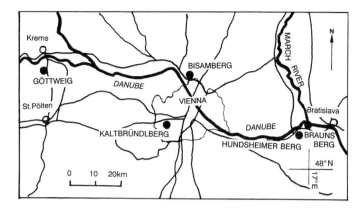

Fig. 3.13. The region around Vienna with location of the investigated objects. Modified after Scheidegger 2002a

Results If the individual results are compared with each other, one obtains the general result that the joint strikes correlate with at least one of the morphological directions in the same area, inasmuch as the tectonic origin of joints is beyond question. This confirms that the morphology has been co-designed by tectonics. In detail, there are the following correspondences between strikes/trends:

– at the Kaltbründlberg: joints $98° \pm 04°$ and hill trends $100° \pm 10°$

– at the Bisamberg: joints $91° \pm 14°$ and streams $91° \pm 18°$

– at Göttweig: joints $162° \pm 21°$ and hill trend ($162°$)

– at the Hundsheimer Berg: joints $174° \pm 11°$ and hill trend ($180°$)

– at the Braunsberg: joints $119° \pm 05°$; $37° \pm 11°$ and Danube ($126°$; $30°$).

In the first four instances, the joint strikes correspond with (central) European neotectonic expectations (approximately NS and EW; cf. Sect. 2.2.1.4 and Table 2.1). In the case of the Braunsberg, the joint strikes correlate with the trends of the Danube before and after a sharp corner, but not with the "European" joint orientations; they probably correspond to an earlier tectonic phase which created the faults now followed by the river. In any case, the "inselbergs" have been caused by tectonics: primarily by recent, but occasionally also by earlier tectonic activity.

3.5.3.3
Inselbergs in Switzerland

General Remarks A series of isolated hillocks have also been investigated in Switzerland. On the whole, they show similar features as those in Austria: The joints conform generally to the local morphological elements; in most cases,

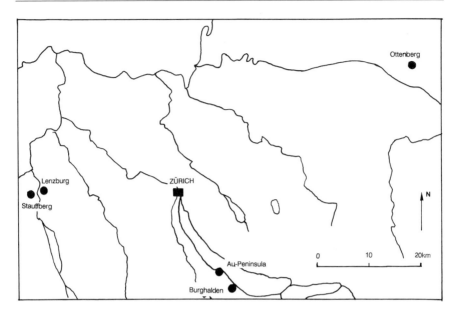

Fig. 3.14. Map of locations of inselbergs investigated in Switzerland

they also follow the general central European pattern. Here we show some examples (location map in Fig. 3.14).

Lenzburg Region   Lenzburg Castle is located atop a hill in the alluvial plain of a small river in the Canton of Aargau; it consists of Tertiary molasse. The orientations of 85 joints were measured at 4 outcrops of the castle hill. They were evaluated statistically according to the usual method (Sect. 1.3). The joint strikes had preferred directions of $74° \pm 10°$ and $159° \pm 06°$ with an intermediate angle of $85°$ and bisectrices at $116°$ and $26°$; the hill trends $66°$. Thus, one joint set strikes parallel to the hill crest, the second orthogonal thereto. All joints taken together yield the normal values for Central Europe.

Stauffberg   Another isolated hill in the Canton of Aargau is the Stauffberg, located SW of the castle hill of the Lenzburg. It consists of molasse composed mostly of marly sandstone, which, nevertheless, occasionally shows outcrops with jointing. It was possible to measure the orientations of 21 joints; the usual statistical evaluation yielded joint maxima striking $85° \pm 11°$ and $171° \pm 18°$ with an intermediate angle of $86°$ and bisectrices trending $38°$ and $128°$. These values correspond to those commonly found in Central Europe.

Ottenberg   A further prominent isolated hill is the Ottenberg in the Canton of Thurgau (Hantke et al. 2003). It consists again of molasse. Joint orientations were measured at 12 outcrops all around the hill, 332 joint orientations were measured at 17 outcrops. The statistical evaluations yielded clearly two well-defined maxima with strikes of $2° \pm 00°$ and $92° \pm 01°$, the bisectrices trending $137°$ and $47°$. The joint strikes can be compared with the river trends in the vicinity; one finds trend maxima at $3° \pm 09°$ and $82° \pm 03°$ with an angle of $89°$ in between and bisectrices at $133°$ and $43°$. Thus, the joint strike maxima are almost identical to the river trend maxima which points to a common origin. These directions also correspond exactly to the "normal" orientations of the joints in Switzerland.

Au Peninsula in Lake Zuerich A prominent landspit protrudes into Lake Zuerich, near the middle of its left (SW) bank. The material consists of consolidated gravels, which, nevertheless, show good jointing. The directions of a total of 78 joints were measured and evaluated statistically according to the usual method (Sect. 1.3). It was discovered that the strike maxima of the joint orientations are well-defined ($124° \pm 15°$ and $43° \pm 17°$), but they do not fit the usual European joint directions (Hantke and Scheidegger 1997). This is an exception to the usual conditions, pointing to a local anomaly that must be present in this region inasmuch as one of the joint strike maxima still coincides with the trend of the crest ($112°$) of the Peninsula. This points to a conformity of the joint direction with the local geomorphological forms and local tectonic stresses.

3.5.4
Escarpent Outliers

Morphology A long, continuously rocky and partly cliffed escarpment ("Niagara Escarpment"), located between the levels of Lake Erie and Lake Ontario, winds its way from the Niagara Falls, Canada, at least to Tobermory on the Bruce Peninsula (Fig. 3.15; Tovell 1979). In the central part of the Escarpment, Mono Cliffs Provincial Park has been established near the Village of Mono Center; the general location of the area is shown in Fig. 3.15; a detailed map is given in Fig. 3.16. In the area of Mono Cliffs, the escarpment itself is skirted by "outliers" which are veritable "inselbergs"; their trend line and that of the valley in between strike roughly N165°E (cf. Fig. 3.16). During the last ice age, a glacier was jammed against the gap between the outliers. The top of the escarpment (Lake Erie level) is flat and swampy and includes lakes.

Origin of the Niagara Escarpment The basic genesis of the Niagara Escarpment is inherent in the cuesta-type of landscape characteristic of SW-Ontario (Lattman 1968): The bedrock consists at the Lake Erie level of gently SW-dipping (ca. $7°$) Silurian strata covered by Late Pleistocene glacial sediments in which differential erosion occurs. This is underlain at the Lake Ontario level by Ordovician shales. The Niagara Escarpment is an exposed erosional scarp in the Silurian strata between the Lake Erie and Lake Ontario levels: Its rim is formed by Middle Silurian dolostones (Johnson et al.1992); the latter are underlain by clays of the lowermost Silurian formation which are eroded rapidly, giving rise to the snake-like escarpment reaching over hundreds of kilometers.

Procedure Turning to the outliers, it is our contention that the valley between the escarpment and the outliers has been predesigned by tectonic processes. As usual, a test of this view can be obtained by regional studies of joints. In the present context, the procedure consisted in a comparison of the joints on the

Fig. 3.15. General map of the Niagara Escarpment, indicating the location of the study area (modifed
after Tovell, 1979)

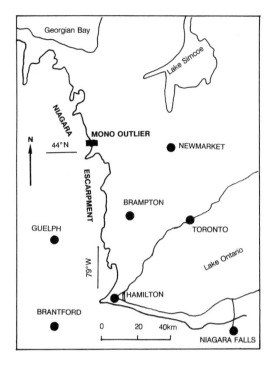

Fig. 3.16. Detailed map of the Mono
Cliffs area, showing the main cliff
and the outlier studied. Outcrops
where joint orientation measurements were made, are indicated by
black dots

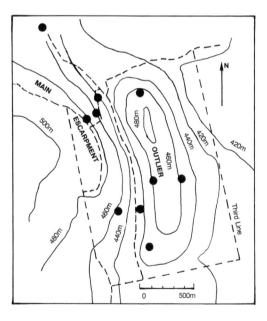

main escarpment and on the outlier shown in Fig. 3.16. Thus, joint orientation measurements were made in the area: The measurements were made at five locations on the main escarpment (Fig. 3.16); these measurements were then compared with measurements made at five locations on the outlier.

Results   The statistical evaluations were made according to the usual statistical method (Sect. 1.3) for the outlier, for the escarpment (cliffs) and for all values combined. Some en-echelon cracks in a trail were noted and their direction was estimated (one single rough estimate). In addition, for a general reference to the surrounding area, the data in the bedrock of Southern Ontario as exemplified by the nearby Rouge River region (from Table 3.4) were also considered. The results are given in Table 3.6 as strike directions for the various groups.

Connection with tectonic stresses   As noted in Sect. 1.4, large-scale observations usually indicate that such joints fit the shear lines rather than the prinicpal stress directions of the neotectonic stress field. Then, the bisectrices of the joint sets would indicate the principal stress directions of the regional stress field. Based upon such a view, the trend of the lines connecting the outliers and that of the valley between the main escarpment and the outliers taken from the map in Fig. 3.16 (Table 3.6) seem to corelate with one of the bisectrices, presumably representing the principal regional pressure direction, corresponding to a stress relief between the outliers and the main escarpment.

Conclusions   One observes that the joint strikes on the Niagara Escarpment near Mono Cliffs, on the adjacent outliers and in the further environment (Rouge River Valley) are essentially parallel, indicating a common neotectonic predesign in these areas. The directions of en-echelon cracks in a trail also correlate more or less with the strike of one of the joint sets. On the other hand, the trends of the line connecting the two outliers and that of the valley between the escarpment and the outliers seem to correlate with one of the bisectrices of the joint sets which is interpreted as maximum tectonic pressure direction, indicating that the outliers would have split off from the main escarpment, the valley in between of course being subsequently enhanced by hydrological processes.

Table 3.6. Statistical evaluations for Mono Center strike directions

Loc.	No.	Max. 1	Max. 2	Angle	Bisectrices	
Joints						
Outlier	98	118 ± 09	24 ± 11	86	161	71
Escarpement	111	109 ± 00	14 ± 02	85	151	61
All at Mono	209	113 ± 08	20 ± 12	87	157	67
Bedrock Rouge	602	118 ± 02	31 ± 04	87	164	74
Cracks (est.)	1	101				
Trends						
Outliers/valley	3				165	

3.5.5
Periglacial Features

3.5.5.1
Introduction

Next, we examine single hummocks whose form has generally been ascribed to the action of glaciers at their beds. This refers to the genesis of roches moutonnées and drumlins.

3.5.5.2
Roches Moutonnées

General remarks   With regard to morphotectonics the most important glacial hummocks are "roches moutonnées", which are solid mounds found on former hard (ice age) glacier beds, now standing free in the present-day landscape. They are elliptical bodies with long axes oriented in the ice flow direction and have been considered as the result of nonuniform, unstable erosion on a hard glacier bed. Roches moutonnées show a smoothly abraded slope on the upstream side of the ice flow; the lee side falls more steeply. Even here, a step-like series of crags and ledges due to the plucking out of joint blocks is commonly observed (Holmes, 1944, p. 217). Ahnert (1996, p. 290) notes that joints can indeed delimit the faces of any rock mound. However, there is more to it: Since joints have been found to be connected with the tectonic stress field, it is clear that *tectonic processes* play a major role in the predesign of such features; they are not caused by an instability of the external erosion (Hantke, 1991, p. 51). We shall give a specific example of a roche moutonnée.

Example: Castle-Hill of Burghalden   The Castle-Hill of Burghalden is a prominent feature in the landscape of the Canton of Zuerich in Switzerland (map in Fig. 3.17). At its western side stands the castle of Wädenswil, which was to protect the Canton of Zuerich from enemy attacks along the Reidbach Creek. At the east end lies the town of Burghalden, from which vinyards strech up the hill. The substratum consists of Quaternary cemented gravels which were subsequently worked over by glaciers during the latest glaciation; the result was the present hill which is therefore interpreted as a roche moutonnée (Hantke et al. 1967). The cemented gravels show jointing; the orientations of the joints were measured at six outcrops. The results of the statistical evaluations yielded am unequivocal picture: one obtains strike direction maxima for the joints of $155° \pm 0°$ (NNW-SSE) and $82° \pm 2°$ (ca. E-W), both sets correspond to the "European" condition with a maximum compression (one of the bisectrices) direction in N118°E, i. e. about NW-SE. Morphologically, one of the joint strike directions (155°) corresponds to the flow direction of the Reidbach Creek (162°), and the dip direction of the first joint maximum (245°) to the inclination direction of the SW-slope (252°) of the hill. A common neotectonic origin of the joints, the hill and the Reidbach is therefore indicated.

Fig. 3.17. Detailed map of the roche moutonnée near Burghalden

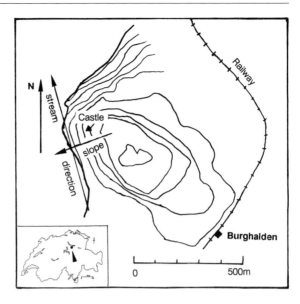

3.5.5.3
Drumlins

The second type of the glacially caused hummocky features mentioned earlier are drumlins. These are swarms of whaleback type hills supposed to have been moulded by flow instabilities in an ice-age glacier flowing over a soft, friable substratum (Holmes 1944, p. 230). Drumlins are thus presumably exogenic features; they are only indirectly connected with tectonics, inasmuch as the glacial flow may have been influenced by preexisting tectonic features. In connection with "morphotectonics" they are of little significance.

3.6
Volcanic Features

3.6.1
Introduction

3.6.1.1
General Remarks

The origin of volcanic landscapes, like that of many other landscapes, has primarily been ascribed to epigenetic processes, in this instance mainly to the "epigenetic" action of lava flows, their cooling and other effects of volcanic eruptions (see e. g. Cotton, 1944, Holmes 1944, 1965). However, tectonic co-designs are quite as prevalent in volcanic landscapes as they are in other landscape types.

To demonstrate this, it is necessary to survey some general aspects of volcanism: notably the types of eruptions and lavas that form the background.

Then, the eruption processes will be investigated with regard to the *mechanical* designs that result from these processes in the landscapes. In this instance, it should be recalled that *mechanical* aspects in a landscape express themselves by their non-randomness, as compared with epigenetic features: This is a consequence of the Principle of Antagonism in landscape evolution (cf. Sect. 1.2.2.2). The mainstay of our argument is focussed on *orientation structures* (of joints, river segments, earthquake fault planes etc.), which will be analysed statistically by the method described in Sect. 1.3. Moreover, joints can generally been considered as parallel to the shear lines in the stress field that generated them (cf. Sect. 1.4.1): Then, the principal axes of the latter (azimuth $N > E$/plunge angle) can be calculated as the bisectrices between the two preferred joint strikes.

3.6.1.2
Common Features

Some background concepts are essential in any discussion of volcanic landscapes. The first of these concern the nature of the *lavas*: they can range from basaltic (45 – 55% silica content) to andesitic (55 – 65% silica content); in between are the "mixed" or "other" types of volcanism (Ollier, 1981). Generally, basaltic lavas are more fluid (less viscous) than andesitic ones, but their chemical composition is not the only determining factor; dissolved-gas content also plays a significant role.

Composition and gas content of the lava greatly affect the *form* of the ejecta issuing from a volcano. Very fluid lavas form large flows that sometimes cover vast regions. The internal dynamics of such lava flows have been investigated on many occasions. Thus, they have been treated as kinematic waves (Baloga, 1987) in a rheological material (McBirney and Murase, 1984), as a Bingham plastic fluid flowing under the action of gravity (Park and Iversen, 1984) or as a gravity-driven slow creeping flow (Hutter and Vulliet, 1985). Their surface may be flat or ropy (pahoehoe lavas); sometimes the congelation leads to blocks (aa lava). When ropy fluid lava erupts at the sea floor, characteristic pillow structures may result. Columnar structures develop inside large lava masses that consolidate under stagnant conditions (Reiter et al., 1987; DeGraff and Aydin, 1987; Aydin and DeGraff, 1988). More viscous lavas containing high amounts of gases become fragmented during generally explosive eruptions; they form pyroclasts. Large fragments are called bombs, small ones lapilli, and very fine ones ash. As these accumulate on the ground and become consolidated, they form volcanic breccias and tuffs. The deposits resulting from glowing particulate clouds (nuées ardentes) are ignimbrites or ash-flow tuffs (cf. Walker, 1983).

Volcanic eruptions lead to characteristic *structures*. Holmes (1944) has classified such structures as explosive (vents, ash and cinder cones), mixed (composite cones of pyroclasts and lavas) and effusive (lava shields, lava plateaus, domes and needles). Which of these structures develop depends on the fluidity of the lava. When a lava chamber beneath a vent is emptied by eruptive processes, it may collapse and give rise to a caldera (Wood, 1984; Druitt and

Sparks, 1984); sometimes the latter resurges by some rebounding process after its initial formation (Marsh, 1984). Gas bubbles in the lava may lead to the genesis of caves (Stein and Pflug, 1985).

Often, water-borne mass movements occur as secondary effects triggered by a volcanic eruption: these are "debris avalanches" or "lahars". The water may originate from an emptying crater lake, a summit ice cap melting instantly in the heat of an eruption, the swapping over of a nearby lake filled by volcanic ejecta, or simply from meteoric processes. The transport and the deposition of the debris are sedimentological processes; the corresponding literature is large: some of it concerns the aqueous transport mechanism (Davies, 1985), most of it the resulting sedimentary structures (e. g. Allen, 1984; Smith, 1986; Pflueger and Seilacher, 1991), and very little of it the clast orientations (Rust, 1972; Kohlbeck et al., 1994).

3.6.1.3
Eruption Types

Corresponding to the chemical composition and gas content of the lava, different *types* of eruptions can occur, each of which give rise to different characteristic volcanic landscape forms. The classification tables are usually given in the form of "one-dimensional" lists (cf. Holmes, 1944, 1965; Poldervaart, 1971; Ollier, 1981); thus lavas range in their composition from basaltic through mixed to acidic (andesitic), which lead to *effusive, mixed* and *explosive* geomorphic forms, respectively. This review is intended to be a discussion of the mechanical designs found in these geomorphic forms with examples from nature, concentrating on cases that the author has investigated himself.

3.6.2
Basaltic/Effusive Volcanism

3.6.2.1
General Remarks

We begin our review of mechanical designs in volcanic landscapes with *basaltic* volcanism. Inasmuch as basaltic lavas are essentially quite fluid, they give rise mainly to structures that have been classified as *effusive* above.

3.6.2.2
Deccan Traps

The Deccan Traps are a great volcanic formation of India that had been emplaced towards the end of the Cretaceous when a large part of the Peninsula was affected by large volcanic eruptions. Whilst the stepped and dissected morphology of the Deccan Traps has mostly been considered as due to erosion, it is evident that such features as the Western Ghats have a geotectonic origin. The studies of this region have already been reviewed in Sect. 2.3.2.3 in connection with the global morphotectonics of Peninsular India. It is evident that joint orientations inside and outside the Deccan Traps, river segment orientations near Koyna and the in-situ measurements in the Kolar gold fields all fit together with the results from the fault plane solution of an earthquake near Koyna. Thus, it is evident that there is no difference between Deccan and non-Deccan regions. This means, as noted, that the patterns in question are caused by tectonic and not by intrinsic lava-genetic or exogenic processes.

3.6.2.3
Oceanic Islands

Similar situations have been found for various basaltic oceanic islands: in the Atlantic for the Macaronesian Islands, in the Indian Ocean for the Mascarenes, and in the Pacific for Easter Island (Isla de Pascua), for some Society Islands (Tahiti, Moorea and Huahine), for Hawaii, and for Samoa, which are all of basaltic volcanic origin: The joints in the lava flows do usually not have a flow origin, but show strike directions corresponding to the respective global neotectonic stress fields. Details have already been described in Sect. 2.4 (on oceans) of this book.

3.6.3
Mixed Volcanism

3.6.3.1
Description

Next, we shall consider some types of volcanism that cannot be considered as basaltic, but not as andesitic either. Many of the Pacific Rim island volcanoes, as well as those in the African Rift Valley, belong to this category; their lavas tend to be relatively rich in alkalis, central eruption vents tend to be dominant (Ollier, 1981). Phenomenologically similar volcanism also occurs in other places of the world. Again, we shall consider some specific cases.

3.6.3.2
Philippines

The morphotectonic conditions in the Philippines have already been reviewed in Sect. 2.2.2.5, where it has been shown that the area is mainly plutonic volcanic and sedimentary in origin dating from the Cretaceous to the Quaternary. The present-day plate dynamics is determined by crustal underthrusting along several subduction systems: underthrusting must occur in a direction more or less normal to the strike (roughly N145°E) of the Philippine archipelago. Thus, one would expect a compression direction of N55°E. Indeed, the interpretation of joint orientation measurements in the area is that one of the bisectrices (N58°E) obtained from the joints corresponds to the plate-tectonic underthrusting direction (N55°E) as explained above. The orientation pattern of the segments of the Paygwan drainage basin agrees with that of the joints; the details are in Sect. 2.2.2.5. Thus, there is a correspondence between river segment and joint orientations in the area: both are supposedly the result of the plate tectonic subduction from the NE. This indicates that the morphology is determined by the regional tectonics and not by the volcanism in the area.

3.6.3.3
East African Rift Valley

Mixed basaltic–andesitic volcanism has also been recognized in the East African Rift valley: Many lavas are basaltic, but silicic volcanics are also common. In the central Ethiopian Rift, the eruptions have produced only few basalt flows, but mainly trachytes and ignimbrites issuing from explosive silicic centers; the volcanism typically has started some Ma ago, culminating about 0.2 Ma ago, but continuing to the present day (see Sect. 2.3.1.3). Joint orientations have been measured in the Rift valley: If their genesis were lava-exogenic, i.e. connected with the emplacement of the lava, they should have a random orientation. This is, however, not the case. Moreover, these orientations agree with those determined for Ethiopia generally (details in Sect. 2.3.1.3) and the bisectrices of the joint strikes are parallel and normal to the rift trends. This can be interpreted as the maximum pressure direction P as parallel and the minimum compression (greatest tension) direction T normal to the strike of the rift and of the faults, paralleling the latter. Thus, it appears as established that the joints in these volcanic outcrops in the Rift are not cooling cracks of lava-"exogenic" origin, but have been mechanically predesigned by the (neo)tectonic stress field prevailing in the area.

3.6.4
Explosive-Andesitic Volcanism

3.6.4.1
General Remarks

Finally, we look at andesitic volcanism. This type of activity is characterized by highly viscous andesitic lavas that are prone to erupting under explosive conditions. There are several types of such volcanism, depending on the pressure obtained in the dissolved gases. We shall discuss a paradigm case in the Caribbean.

3.6.4.2
Caribbean Islands

As a paradigm, we take the island of Guadeloupe, and specifically its volcanic Basse-Terre part with its Soufrière. A specific study of some of its geomechanical aspects has been made by Bonneton and Scheidegger (1982). A set of faults and fissures had been identified by Westercamp and Tomblin (1979) on Basse

Terre whose orientation structure was analyzed according to the methods explained in Sect. 1.3 and compared with a corresponding analysis of 614 joint orientations made by Bonneton and Scheidegger (1982). The results were for the strike-azimuths (degrees N > E):

– Faults Max. 1: 141 ± 10 Max. 2: 53 ± 10

– Joints Max. 1: 157 ± 11 Max. 2: 71 ± 09

Evidently, both maxima agree between faults and joints within their statistical error limits; hence it can be stated that they have presumably the same origin. Furthermore, Bonneton and Scheidegger (1982) have shown that the joints on Guadeloupe correspond to those expected from plate-tectonic stresses in the area; hence the faults around the Soufriere identified by Westercamp and Tomblin (1979) are postulated to be of the same origin: They are not connected with the volcanism per se, but have been predesigned by geotectonic processes.

3.6.5
Conclusions

Reviewing the discussion in this section, we can state that many (if not most) surface features in all three types (effusive-basaltic, mixed, and explosive-andesitic) of volcanic areas are not essentially due to "epigenetic" volcanic activity, but have been predesigned substantially by neotectonic conditions. This refers not only to the valleys, but also to such features as fissures and lineaments. The proof for this contention lies in a comparison of the orientation structures of the mentioned features to that of the joints in the respective areas, which have been shown above to correspond to and be caused by the neotectonic stress field the world over.

3.7
Mass Movements on Slopes

3.7.1
General Remarks

Surface mass movements cannot be considered separately from landscape development. The latter has been shown (Sect. 1.2.2) to be governed by a series of principles of which the Principle of Antagonism is the most important: A landscape is the instantaneous result of the action of two antagonistic processes: the endogenic buildup and the exogenic denudation. Since one encounters tectonic elevation rates of the order of mm/a (km/Ma) in mountainous areas and the mountains have not become much higher in the last few million years (Ma), a substantial "flow" of mass must take place constantly over the landscape surface: This "flow" is represented by the surface mass movements, particularly slides, which, indeed, are part and parcel of the normal landscape development.

Regarding all sorts of mass movements (slow slumps, debris flows, land-slides; for a general qualitative characterization see e.g. Dikau et al.1996, p. 1–12) one can make the following general statements:

The mass movements (i) *occur in spurts*, they are part of the *normal landscape development* (their incidence is a stochastic time series with a fractal structure) and (ii) *their direction is predesigned* by the neo-tectonic stress field; the latter also generates the *orientation* of the joints, hence the direction of the joints and the slide motion directions are correlated.

Thus, whilst the immediate *triggering* of mass movements on slopes is undoubtedly due to external (meteorologic or seismic) processes, their *geometry* is primarily designed by the tectonic stress field. The direction of the mass movements is along the inclination of the valley sides. If the valleys are of a recent origin, they are parallel to the joint strikes (see above); thus most slides originate in parallelepipedal fracture niches. This is not the case in older valleys, especially those which originated during the emplacement of nappes: in this case, mountain fractures and valley closures are the result: the landslides occur as stress relief fractures at the tear scars, which are oriented at right angles to a principal tectonic stress direction. Incidentally, the same is true for artificial cuts. Thus, there exist two types of mass movements on slopes: (1) wedge/shear motions and (2) mountain fractures/valley closures (Ai and Scheidegger 1984b).

The wedge-shear motions are probably the most frequent. They are characterized by the fact that the motions are parallel to one of the joint sets. They arise in a fracture niche and have the form of a wedge. They are usually encountered in tectonically active regions, such as in the Alps or in the Himalaya.

The mountain fractures/valley closures occur when a valley or an artificial cut was formed independently of the neotectonic conditions: In this case, the mass movements occur primarily on slopes which are oriented at right angles to a principal tectonic stress direction; the latter may be the direction of the largest (P) or the smallest (T) compression. The typical natural example hereof is the "mountain fracture", in which a "graben" (fissure) is formed at the crest of a ledge and a slumphump at its foot. Specific case studies of the two types of slides will be discussed below.

3.7.2
Shear Slides

3.7.2.1
General Remarks

Shear or wedge slides are the most common type of slides. They are characterized by a displacement direction at their surface which is parallel to one of the joint directions in the area. Although, like all slides, they occur primarily in neotectonically active regions such as the Alps and the Himalayas, they may also be found as more local phenomena in tectonically rather quiet places. We shall discuss some of these in detail.

3.7.2.2
Shear Slides in the Alps

a. Shear slides in Switzerland

Switzerland, as a classical Alpine country, offers many examples of slides of every size and character, including shear slides. Figure 3.18 shows the location of the objects discussed here.

Thus, between the the villages of *Sattel and Schwyz* (Ct. Schwyz, Switzerland) there is a slide area in glacially reworked molasse (Buser and Scheidegger 1992). The sliding mass consists of argillaceous-arenaceous slope clay. Exact observations during several years show that the slide is a classical example of a slump (Fellenius 1927): In consequence of consolidation and sliding, a semi-circular niche is created in the head area of the mass movement; at the same time, a slumphump is created at the foot. A set of marker points were established and the displacements that occurred during ten years were determined by repeated aerial photogrammetry. Also, a stable mass was found to have slid downhill 10 m during that time. In the general area surrounding the slide, joint orientation measurements were made at four locations, so that it was possible to relate the slide morphology to the tectonics of the area. The data for the joints and for the displacements were evaluated statistically according to the standard method (Sect. 1.3.3). It turns out that the displacement direction closely parallels one of the joint strike directions (see Table 3.7). This is indicative of failure along the shear lines of the ambient tectonic stress field.

A further (actually: ongoing) slide is situated near *Schinznach* in the Canton of Aargau at the eastern edge of the Swiss Jura (Gerber and Scheidegger 1984). The slide material consists of opaline clay of the lowest Dogger formation (middle Jurassic). The slide has been operative continuously at least since the end of the nineteenth century; detailed studies were made by using maps from the years of 1910 and 1924, and by direct observations between 1960 and 1978. It was found that the 200 m-long slide with an inclination of 16°, is active in spurts: Short periods of activity alternate with periods of quiescence.

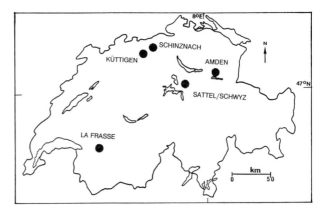

Fig. 3.18. Location of the slides in Switzerland

Table 3.7. Summary shear slides, strikes/trends

Feature	No.	Max. 1	Max. 2	Angle	Bisectrices	
Alps						
Switzerland						
Sattel-Schwyz						
Joints	85	169 ± 12	66 ± 13	76	28	117
Displacements	5		68 ± 04			
Schinznach						
Joints	82	103 ± 16	11 ± 13	89	146	58
Displacements	16		22 ± 00			
Kuettigen						
Joints	63	166 ± 21	88 ± 17	78	127	37
Fissures	5	162 ± 06				
Displacements	2	109				
Austria						
Semmering						
Joints	66	137 ± 13	43 ± 14	86	0	90
Displacements	30		57 ± 00			
Woerschachwald						
Joints						
Within slope	29	117 ± 20	36 ± 10	71	75	167
Surround. frame	424	161 ± 12	81 ± 16	80	121	31
All joints	453	159 ± 10	78 ± 16	81	118	28
Displacements	2	167 ± 04				
Bad Gastein						
Joints	413	126 ± 00	24 ± 02	77	165	75
Displacements	239	132 ± 04				
Himalaya						
Garhwal Himalaya						
Kaliasur Slide						
Joints	36	157 ± 12	65 ± 29	88	21	111
Displacements	16	157 ± 19				
Chinese Himalaya						
Sale Shan Slide						
Joints	234	94 ± 03	163 ± 06	69	127	39
Displacements	3		178 ± 26			
Modest Hills						
West Africa						
Ankpa						
Joints	19	103 ± 13	191 ± 06	88	146	57
Gullies	8	105 ± 19				
Displacements	8		195 ± 19			

Fig. 3.19. Detailed map of the slide near Küttigen, Ctn. Aargau, Switzerland

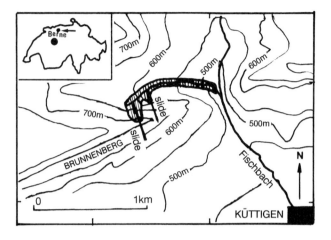

A correlation study showed that heavy precipitation causes the slide to move at the head after 1 day, in the middle after 3–4 days and at the foot after 10–12 days. Apart from the kinematics, the morphotectonics of the slide is of interest. For this purpose, joint orientation measurements were made and then evaluated according to the statistical method outlined in Sect. 1.3.3; the results are shown in Table 3.7. The directions of the displacements were taken as parallel to the edges of the slide; since the latter are turning somewhat, several sections were measured individually and the whole evaluated according to the standard statistical method (Sect. 1.3.3). The results are also summarized in Table 3.7. It is observable that the slide direction is (within error limits) parallel to one of the joint sets, which is indicative of a shearing motion.

Another slide in the Swiss Jura Mountains occurred during the night of 23/24 February 1999 near *Kuettigen* in the Canton of Aargau. The slide(s) originated from the crest of the Brunnenberg (cf. Fig. 3.19); there seemed to have been two events, inasmuch as two slide scars can be clearly identified. The further mud-flow down the tributary of the "Fischbach" was an independent consequence of the slides. A report on the event was published by Eberhardt (1999) from which the motion directions (around N20°W-N18°W) could be deduced morphologically from the outline of the slide edges (Fig. 3.19). During excursions led by Dr. Eberhardt, the author was also able to measure the orientations of some joints (at 3 locations, one at the valley bottom and 2 at the crest of the Brunnenberg) and of the fissures that were visible along the crest. The orientation measurements were statistically evaluated; the results are shown in Table 3.7. Thus, it was possible to recognize the slide as a shear slide; the strikes of the joint orientations fit the motion directions and also fit those of (northern) Switzerland (156° and 70°, cf. Table 2.1).

b. Shear slides in Austria

Austria is another classical Alpine country in which many slides have been studied. A detailed description of a series of shearing slides in this country has been given by Scheidegger (2000a). We give here a summary of these studies (location map in Fig. 3.20).

Thus, near the *Semmering Pass* an unstable area exists. Geologically, it is situated in the lower Austroalpine formation consisting of Mesozoic sequences, mainly of Keuper, an argillaceous material that becomes plastic when wet. The slide was kept under observation for several years (Castillo 1997). The motion along a profile line was geodetically determined during one year and corresponding displacement vectors were measured and compared with joint orientation measurements in the same area. The results of the evaluation of the measurements (Table 3.7) made it clear that the slide motion does not occur in the line of steepest descent, but that it coincides with the strike direction of one of the joint strike maxima. Thus, one is faced with a shearing motion.

On the north side of a small valley (*Woerschachwald*) running parallel to the Enns river near Liezen in Styria (location map in Fig. 3.20), there is a large slope in slow (fast during spurts) motion (Hauswirth et al. 1982). Geologically, the slide lies entirely in Upper Cretaceous Gosau-deposits which are characterized by sequences of coarsely clastic conglomerates and sandstones. Morphological evidence for mass movements is ubiquitous: Slumphumps, bent trees and tear scars are seen everywhere. Therefore, a geodetic net of benchmarks was set up and remeasured three years later during which displacements of 15–17 cm with an azimuth of $167° \pm 04°$ were found to have occurred. These were compared with joint orientations in the area (Table 3.7); the results of orientation measurements *outside* the sliding mass yielded strike azimuths of $161° \pm 12°$ and $81° \pm 16°$, but quite different values *within* the slide ($117° \pm 20°$ and $36° \pm 10$). This shows that the slide as a whole follows the shear lines of the ambient neotectonic stress field, but the sliding mass becomes internally broken up: One is faced with a shear slide breaking up within by local stress relief.

Mass movements had also been observed in the town of *Bad Gastein* (location map in Fig. 3.20) for a long time. Inasmuch as cracks and settling phenomena occur, they express themselves in damage to buildings. Geodetic measurements showed that movements were in part 80 cm in 41 years (Hauswirth and Scheidegger 1980, Figdor et al. 1990); their directions were

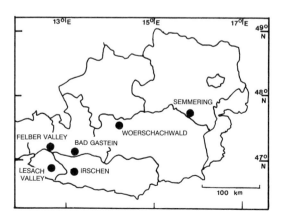

Fig. 3.20. Location map of the slides in Austria (modified after Scheidegger 2000a)

around 132°. These were compared with the results of joint orientation measurements which yielded two sets of strike directions (see Table 3.7). There is obviously an agreement between one of the preferred joint strikes and the mean displacement direction. Thus, one can infer that the slide follows one of the shear lines of the stress field and that the observed displacements are definitely of tectonic design.

3.7.2.3
Shear Slides in the Himalaya

Garhwal Himalaya: Kaliasaur Slide (Uttar Pradesh, India)   In the Garhwal Himalaya (field guide by Sah and Bist 1998), a morphotectonic study of a slide near the village of Kaliasaur between Srinagar and Rudraprayag was made (map in Fig. 3.21): joint orientations were measured on outcrops at both sides of this slide; their values were compared with the general displacement directions of earth movements in the area given by Sah and Bist (1998); the orientations of these directions can be treated like joint strikes. There is really only *one* numerically calculable maximum (Table 3.7). A comparison of the joint orientations with the displacement directions shows that the joint strikes and the displacement directions are parallel (Scheidegger 1999a). This is a condition characteristic of slides caused by shear- or wedge-type failures.

Fig. 3.21. Location of the Kaliasaur slide in the Garhwal Himalaya. *Open circles*: locations of joint orientation measurements. Modified after Scheidegger (1999a)

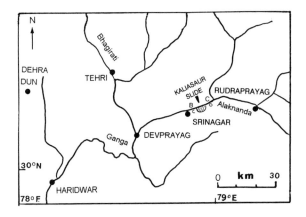

Chinese Himalaya: Sale Shan Landslide (Gansu, China)   A catastrophic landslide occurred (Ai and Scheidegger 1984b) on 7 March 1983 at Mt. Sale (Sale Shan) in southern Gansu, China (103°35′10″E; 35°33′40″N; cf. location map in Fig. 3.22) which killed 220 people. Geologically, the bedrock in the surrounding area was clay rock of the late Neogene; it was covered by lithic and aeolian loess of the Quaternary. The slide consisted of two spurts 65.5 seconds apart. The mean motion direction (calculated from Fig. 1 in Ai and Scheidegger 1984b) of the slide was in the direction of N178°E; the slide scar was straight and smooth striking N80°E (Table 3.7). The displacement direction was compared with

Fig. 3.22. Locations of slide areas in China discussed in this book

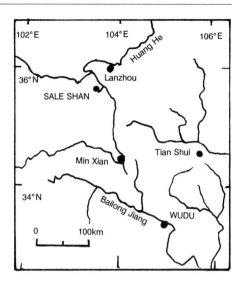

joint orientations measured in the vicinity of the slide; the results of a statistical evaluation (method in Sect. 1.3.3) are shown in Table 3.7. These results are consistent with the hypothesis that the slide is tectonically predesigned: Its motion direction follows one of the shear lines of the neotectonic stress field; the slide scar is parallel to the other.

3.7.2.4
Shear Slides in Modest Hills

Mass movements do not only occur in mountain areas, but also in tectonically rather quiet hills such as those near *Ankpa (Fig. 3.23) in the SW part of Benue State of Nigeria* (Scheidegger and Ajakaiye 1994). Because of the lack of much relief, the causes of the mass movements have generally been sought in the action of meteoric water. It has actually been possible to predict the occurrence of individual events based on such considerations (Okagbue 1989). Nevertheless, the instabilities in the ground have a much deeper primary cause as well; they are connected with the prevailing neotectonic conditions.

The area (Fig. 3.23) is characterized by laterite-covered hills; the laterite overlies siltstones and unconsolidated sands from the Cretaceous. The hills are cut by numerous gullies and form (sub-)parallel ridges which strike from E to W. They are about 100 m high and appear to be essentially erosion remnants. The laterite forms a hard crust ($\sim$ 1 m thick) which acts as a protection of the rather friable material below, so that the top of the hills is generally quite flat. The primary unstable features appear to be the gullies. Slides and creeping phenomena occur ubiquitously on the banks of the creeks and of the main rivers. It is therefore the formation and evolution of the creeks which is the primary process in the origination of mass movements; the actual mass-movements would then be at right angles to the gullies, on the slopes of the latter.

Fig. 3.23. Location of Ankpa in Benue State, Nigeria. *Black dots* – towns; *thin lines* – rivers; *thick lines* – main roads. (Modified after Scheidegger and Ajakaiye 1994)

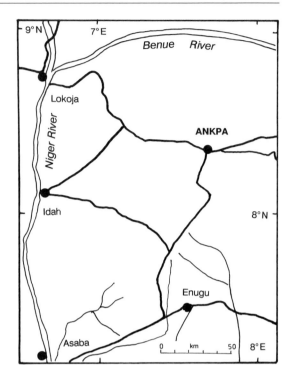

The gullies are seen to form a definite orientation pattern, which points to a tectonic design. This contention is supported by the fact that cracks appear in the ground ahead of the gullies. The orientation pattern of eight gullies and cracks has been determined statistically according to the method described in Sect. 1.3.3: there is, in fact, only one direction maximum present (Table 3.7). The displacements, as noted, are then at 90° to this direction. This result was compared with joint orientation measurements made in the laterite around the gullies; there are evidently two identifiable maxima (Table 3.7). A comparison of the two types of features indicates that gullies, cracks and joints are basically similarly oriented and have therefore probably been generated by the same cause, viz. by the action of the *neo*tectonic stress field (since the joints had been found in *recent* laterite. The actual mass movements are then simply slumps down the slopes of the gully walls and are thus also basically *designed* by the neotectonic stress field, although they may have been *triggered* by hydrological effects.

3.7.3
Slides at Mountain – Fractures and Artificial Cuts

3.7.3.1
General Remarks

Thus far, we have considered shear slides. However, as noted earlier, if a valley or a cut has been created *independently* of the neotectonic conditions, slides will preferentially occur on faces oriented at right angles to one of the principal tectonic stress directions represented by the bisectrices of the joint sets – this, incidentally, can be the tension (T)- or the compression (P)-direction: The typical case in nature occurs on the form of a "mountain fracture", where a "tensional fracture" (in form of a ditch) at the ledge of a ridge is opening up, but at the bottom a slumphump is formed which stands under pressure: This represents the classic morphology of a "mountain fracture" above a "valley closure" below. Again, mountain fracture slides occur mainly where mountains are. Thus, we shall discuss some of such cases following the same order as with shear slides: Slides in the Alps (Switzerland, Austria) and elsewhere (China).

3.7.3.2
Mountain Fractures in the Alps

a. Switzerland   A land slide occurred to the west of *Amden* on the north shore of the Walensee (Canton of Saint Gall; location see Fig. 3.18; detailed map in Fig. 3.24) on 21 January 1974. Geologically, the ground consists of various types of Cretaceous limestones which are heavily jointed and can break into single large masses. At the lake shore there was at that time an active limestone quarry. The direction of the slide was essentially downhill, i.e. N192°E; the slide scar strikes about N110°E. This was compared to joint orientation measurements in the area (Scheidegger 1985a); these were statistically evaluated according to the usual (Sect. 1.3.3) method; the results are shown in Table 3.8. An inspection of this table shows at once that the displacements are not parallel to the joint strikes, but rather to the bisectrices of the latter. This is typical of externally induced slides: The direction of the slide is downslope to the lake; this direction is not of a neotectonic origin, but is connected with the pre-Miocene emplacements of the Alpine nappes; the slide may well have been connected with the quarrying operations at the lake shore. It is well known that valleys or artificial cuts normal to a general tectonic stress direction have a destabilizing influence (Scheidegger and Ai 1986).

A rather important slide occurred at *la Frasse* (near Leysin and Forclaz) in the Canton of Vaud (location in Fig. 3.18), Switzerland (Scheidegger 2000b). Geologically, the slide occured on a river valley slope (of the Grande Eau River) consisting of ultrahelvetic flysch material which abuts against a wall of middle Triassic dolomite on the opposite side of the river; the slide directions were given as equal to N152°E (Noverraz and Bonnard 1990). Joint orientation measurements were made in the area; the results of the statistical evaluations (method of Sect. 1.3.3) were compared with the slide displacement vectors as

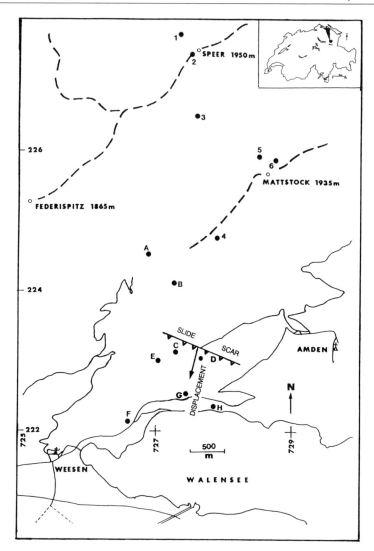

Fig. 3.24. Sketch map of slide near Amden on the north shore of the Walensee

shown in Table 3.7. It will be acknowledged that the slide motion does not agree with the joint strikes, but agrees within 1° with the direction of one of the bisectrices, which presumably represents the maximum neotectonic compression direction in the area. This is characteristic of "valley-closure slides".

b. Mountain Fractures in Austria   A review of mountain fracture studies in Austria is included in a general study of Austrian landslides (Scheidegger 2000a); we give a short summary here. Thus, a slide area in the *Felber Valley* (location map Fig. 3.20; in detail in Fig. 3.25) represents a classic example of a mountain fracture: the slide occurs below a fairly straight ledge (between Archenkopf and Brentling; see Fig. 3.25) down to the valley bottom. Detailed tachymetric, seismic, geomorphological and geological studies showed (Carniel et al., 1975) that the top of the the ledge contains morphologically characteristic features of mountain fractures, such as opening fissures, which are compensated at the valley bottom by valley closure: slow movements occur from the ledge downward to the valley floor. The motion directions can be assumed to be at right angles to the mountain fissures at the top of the ledge. The directions of 14 of these have been measured and statistically evaluated according to the procedure given in Sect. 1.3.3. The results are shown in Table 3.8; the slope movements (also shown in the Table) are normal to the fissure trends. For a comparison with neotectonics, the orientations of joints were measured at six locations (see map in Fig. 3.25) were measured. The maxima referring to the steeply dipping sets are shown in Table 3.8. It turns out that the directions of the motions, being normal to those of the fissures, appear to be correlated (about within the the error limits) with that of one of the bisectrices

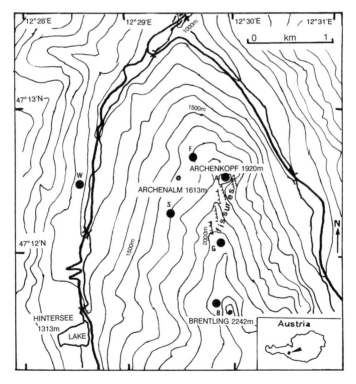

Fig. 3.25. Sketch map of the mountain fracture in the Felber Valley in Salzburg Province, Austria. Locations of joint orientation measurements are shown as *lettered black dots*, isohypses (100 m equidistance) as *thin lines*, rivers/lakes as *medium heavy lines* and roads/highways as *thick lines*

(presumably the P direction) obtained from the joint orientations, indicating a push corresponding to a "valley closure" at right angles to the mountain fractures.

Further deep-seated mass movements in Austria have been investigated on a slope above the village of *Irschen in Carinthia* (location see Fig. 3.20). The morphology of the region shows features typical of mountain fracture/valley closure again: Fissures and tear scars in the upper part of the slope and slump-phumps in its lower part. Heavy rain falls, particularly when occurring at snowmelt time, seem to lead to increased sliding activity (Scheidegger et al. 1984). For a study of the area, a basic geodetic net with bench marks was set up in 1975 and remeasured four and eight years later; a significant downhill motion was discovered (see Table 3.8). This result was correlated with joint orientation measurements at nine locations which were then statistically evaluated according usual (Sect. 1.3.3) method; the results are given in the Table 3.8. The two main joint strike direction maxima were compared with the displacement direction. It is seen that the latter correlates closely with one of the bisectrices, which is typical for valley closures.

Finally, the *Lesach Valley in East Tyrol* (Hauswirth et al. 1979), Austria (location in Fig. 3.20) is subject to obvious valley closure. Its flanks are steep, grown over by grass and show traces of mass movements everywhere. At the ledge, mountain fracture fissures form the corollary of the valley closure below. The morphology shows movements. Again, a geodetic net with bench marks was set up to monitor the displacements during four years. Displacements were found at most within the slide area; the results are listed in Table 3.8. As usual, the ambient neotectonic conditions were derived from joint orientation measurements at 19 locations by the method explained in Sect. 1.3.3; the results are also shown in Table 3.8. It is seen that the displacement direction maximum closely agrees with one of the bisectrices (principal neotectonic stress directions), which is typical for valley closures. Thus, the displacements are predesigned by the neotectonic stress field (Hauswirth et al. 1979). Furthermore, large displacements are known to have occurred in the years 1965 and 1966, during which severe rain storms triggered devastating debris flows. This indicates again that the really large mass displacements are intermittent and occur in spurts.

3.7.3.3
Landslides and debris flows in the Wudu (Gansu, China) Region

The region around Wudu ($33°23'$N, $104°57'$E) is situated in SE Gansu (general location see Fig. 3.22; details Fig. 3.26) and is subject to rapid geodynamic development: large uplift rates exist on the one hand and large downcutting rates on the other. Because of the intensity of the antagonistic action, the intensity and the frequency of slides and debris-flows is high, and the region belongs to one of the most endangered areas with regard to such accidents in China. By a rough estimate there are more than 1000 debris-flow gullies in the Wudu region and they cover an area of 6400 km^2; of these, the bigger debris-flow gullies are estimated to number 400 (Scheidegger and Ai, 1987).

Table 3.8. Mountain fracture slides, strikes/trends

Feature	No.	Max. 1	Max. 2	Angle	Bisectrices	
Switzerland						
Amden						
Joints	319	147 ± 00	59 ± 00	88	103	13
Displacement	1					12
La Frasse						
Joints	63	110 ± 10	16 ± 11	87	153	63
Displacements	8				152	
Austria						
Felber Valley						
Joints	67	112 ± 12	28 ± 02	83	70	160
Fissures	14	178 ± 15				
Displacements	14				88 ± 15	
Irschen						
Joints	784	68 ± 13	175 ± 13	74	32	122
Displacement	1				33	
Lesach Valley						
Joints	446	159 ± 07	67 ± 07	88	23	113
Displacements	10				28 ± 08	
China						
Wudu Region (Gansu)						
Joints	216	169 ± 09	76 ± 07	87	32	121
Gully trends	122				22 ± 04	
Slide direct.	122					112 ± 04

However, there are indications that the morphology of the slides has also a tectonic component in it. In order to find a possible connection between slides/debris flows around Wudu with the prevailing neotectonic conditions, joint orientation measurements were made in the area which were then evaluated according to the usual (Sect. 1.3.3) method. The results are shown in Table 3.8. These results can be compared with the directions of the major debris-flow gulleys along the Bailong River, schematically shown in Fig. 3.26. The results of the numerical evaluation of these directions by the usual statistical method is also shown in Table 3.8. The slide-displacement directions would, then, be normal to the strike of the gulley walls.

There is, therefore, a rough correspondence between the mean direction of the debris-flow gulleys and one of the principal tectonic stress directions derived from the joint orientations. The numerical discrepancy between the two is ±9°, which is within the error ranges of the joint/gulley-direction maxima. The gulley-walls and therewith also the slide-displacements would therefore also be parallel to a principal tectonic stress directions, which has been found to be condition liable to make the gulley-walls collapse causing the debris flows.

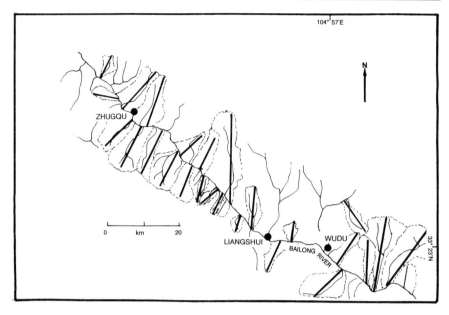

Fig. 3.26. Map of region around Wudu (33°23′N, 104°57′E) situated in SE Gansu showing schematically gulleys (*dotted*) and major debris-flow directions along the Bailong River

3.7.4
Conclusions

Thus, mass movements on slopes are part and parcel of the normal landscape development cycle, about which the following statements can be made:

(i)   The ongoing uplift in a mountainous or hilly region is compensated by mass movements which lead to a quasi-stationary state by a process–response mechanism.

(ii)   The mass movements *occur in spurts*; their incidence is a stochastic time series with a fractal structure.

(iii)   *The direction of mass movements is predesigned* by the neo-tectonic stress field; the latter also generates the *orientation* of the joints, hence the direction of the joints and the slide motion directions are correlated. The direction of the mass movements is naturally in the direction down the slope of the valleys. If the latter are natural, they are parallel to the prevailing joint strikes which are somehow indicative of the shear trajectories of the neotectonic stress field. The landslides, thus are "shear"- or "wedge"-type failures. However, this is not true in the case of older valleys caused e. g. by nappe edges emplaced much before the present-day (then the latter lead to "mountain fracture" and "valley closure"), and particularly by artificial cuts: in such cases, slides occur mainly on faces running at right angles to one of the principal neotectonic stress directions. Evidently, the stability of the object is reduced in this case and slides occur more frequently than if the valleys or cuts run parallel to such principal stress directions. These remarks have been illustrated above by many specific examples.

3.8
Local Morphotectonic Conclusions

Finally we come to a summary of the conclusions referring to the local morphotectonic studies in this book.

Regarding *valleys*, we note that our studies do not support the exclusively erosional origin of river valleys which is todate the practically universally accepted theory: There is plenty of evidence that rivers follow tectonic structures, so that one can speak of a tectonic co-design of many morphological patterns found in riverine landscapes. Thus, drainage orientation patterns in areas that are not completely plane and featureless are to a large extent determined by neotectonic processes. Furthermore, river courses in plan are non-tectonically generated only in completely flat, unstructured plains. Otherwise, some tectonic principle (evidenced by the parallelism of joint strikes and river segments) is active which determines the flow directions. Tectonic events also have a significant influence on the longitudinal profiles of rivers. Compared with equilibrium profiles, the longitudinal profiles of rivers are usually oversteepened in their head water regions and contain stable knickpoints, which can only be explained by the tectonic uplift of the landscape. Similar remarks can be made regarding the transverse cross sections of fluvial valleys which have generally been ascribed to purely intrinsic fluvial processes. Thus, it is usually maintained that the "natural" or "original" (youth) valley forms were V-shaped and have become widened with progressing scouring and steepening of the channel sides by the river itself; particulary glaciers have been thought to be apt to transform V- into U-valleys. However, now there is overwhelming evidence that the erosive action is insufficient for the creation of the valley profiles. Thus, river-genesis must be influenced by the tectonic stress-induced faulting and jointing and the underlying tectonic structure. The same holds true for the supposed overdeepening by fluvial/glacial erosion. Similarly, the genesis of gorges has been ascribed to the erosive action of rivers; but, like in all river valleys, the erosional power is insufficient for the deepening of valleys in solid rock. Thus, tectonic stress-induced faulting and jointing must be the primary reason for the genesis of gorges; the river only erodes preexisting crags. In summary, many valley features are, contrary to the "established" opinion, not primarily caused by exogenic agents, but are to a large extent tectonically co-designed.

Basins are depressions in the landscape. Generally, their origin is obviously old-structural, but they, too, have been affected by *recent* tectonic processes, a fact which expresses itself mainly in a discordance between the tributary river directions and the orientations of such basins: The course of the tributaries has been determined by neotectonics, whilst the layout of the basins is due to old structural conditions.

Turning to *shore and coast lines*, we note again that mainly exogenic processes, such as wave and ice action have been advocated as determinant factors in their genesis. However, neotectonic forces have also had a most important influence. This has been demonstrated by examples from the Lake Ontario shore and the fiords of Greenland and Baffin Land. In many cases, exogenic agents have only reinforced preexisting tectonic features.

The same holds true for "*inselbergs*", isolated hillocks and peaks that are not only common in arid-type landscapes, but also in other types, such as piedmont areas: In glacial landscapes they occur as roches moutonnées and drumlins. Regarding their genesis, exogenic agents (denudation and erosion) have again been assumed as principal causes, but the argument here is that the unusual forms of most of these mounts have been co-designed by tectonics. This has been supported by a comparison of the orientations of the joints on these features with the directions of other geomorphological elements in their vicinity: In most cases the joint strikes correlate with at least one of the morphological directions in the same area. Inasmuch as the tectonic origin of joints is beyond question, this confirms that the morphology has been co-designed by tectonics.

Similarly, the origin of *volcanic landscapes* is not only due to the cooling of lava and other effects of volcanic eruptions: Tectonic co-designs are also present. Thus, joint orientations inside and outside the Deccan Traps are similar and fit together with the results from the fault plane solutions in the area. This means that the patterns in question are caused by tectonic and not by intrinsic lava-genetic processes. Similar situations have been found for various basaltic oceanic islands. Most striking are also the conditions in the East African Rift valley: The joint orientations in the valley agree with those determined for Ethiopia generally and the bisectrices of the joint strikes are parallel and normal to the rift trends. This can be interpreted as the greatest tension direction lying normal to the strike of the rift, as corresponds to the latter's opening-up. In some of the Caribbean Islands, viz. on Guadeloupe, a set of faults and fissures has been identified on its main volcano (Soufrière) whose trends agree with the joint strikes in the area. Furthermore, the joints on Guadeloupe correspond to those expected from stresses in the area suppoed by plate tectonics; hence the faults around the Soufriere are probably not connected with the volcanism per se, but have been predesigned by geotectonic processes.

Finally, it was discovered that the *direction of mass movements* is also pre-designed by the neotectonic stress field: The latter generates the joints whose strikes are correlated with the the slide motion directions. If the slides occur along recently formed valley sides, they are parallel to the prevailing joint strikes which are indicative of the shear trajectories of the neotectonic stress field: The landslides are "shear"- or "wedge"-type failures. This is not true in the case of slides in older valleys caused e. g. by nappe-edges emplaced much before the present day, and particularly slides along artificial cuts: In such cases, slides occur mainly on faces running at right angles to one of the principal neotectonic stress direction.

These remarks conclude our book. By many examples and theoretical arguments we have attempted to demonstrate the importance of (neo)tectonics in geomorphology.

References

Adams J (1980) Contemporary uplift and erosion in the Southern Alps (N.Z.). Bull. Geol. Soc. Am., Pt II, 91(1):1–114

Ahnert F (1996) Einfuehrung in die Geomorphologie. UTB Verlag Ulmer, Stuttgart

Ai NS, Scheidegger AE (1981) Valley trends in Tibet. Geomorph. 25(2):203–212

Ai NS, Scheidegger AE (1984a) The seismotectonic stress field in the Chinese craton. Northwestern Seismolog. J. (Lanzhou) 6(3):1–15

Ai NS, Scheidegger AE (1984b) On the connection between the neotectonic stress field and catastrophic landslides. Proc. Int. Geol. Congr. 27th, Moscow K06:180–184

Ajakaiye DE, Olatinwo MO, Scheidegger AE (1988) A possible earthquake site near Gombe. Bull. Seismolog. Soc. Am. 78:1006–1010

Ajakaiye DE, Scheidegger AE (1989) Joints on the Jos Plateau. J. Afr. Earth Sci. 9(3/4):725–727

Aki K, Richards PG (2002) Quantitative Seismology, Second Ed. University Science Books, Sausalto, California

Allen JRL (1984) Sedimentary Structures: their Character and Physical Basis, Elsevier, Amsterdam, 2 vols

Allis RG (1981) Continental underthrusting beneath the Southern Alps of New Zealand. Geology 9:303–307

Anderson EM (1951) The Dynamics of Faulting and Dyke Formation, 2nd ed. Oliver and Boyd, London

Andrews JT, Buennell GK, Wray JL, Ives JD (1972) An early Tertiary outcrop in north-central Baffin Island, N.W.T., Canada: environment and significance. Canad. J. Earth Sci. 9:233–238

Angelier J (1977) Essai sur la néotectonique de l'arc Égéen et de l'Égée méridionale. Bull. Soc. Géol. France (7)19(3):651–662

Augustithis SS (Ed. committee secr.) (1990) Critical aspects of the plate tectonics theory. Theophrastus Publ. Co, Athens. 2 vols

Aydin A, DeGraff JM (1988) Evolution of polygonal fracture patterns in lava flows. Science 239:471–476

Babcock EA (1975) Fracture phenomena in the Waterways and Fort McMurray formations, Athabasca Oil Sands Region, Northeastern Alberta. Bull. Canad. Petrol. Geology 24:457–470

Bahat D (1991) Tectonofractography. Springer, Berlin, New York

Bak P, Tang C, Wiesenfeld K (1988) Self-organized criticality. Phys. Rev. A38:364–374

Baker BH (1963) Geology and mineral resources of the Seychelles Archipelago (with map 1:50000). Geol. Surv. Kenya Mem. 3:1–140

Baloga S (1987) Lava flows as kinematic waves. J. Geophys. Res.92(B9):9271–9279

Bankwitz P, Schneider G, Kaempf H, Bankwitz E (2003) Structural characteristics of epi-central areas in Central Europe: study case Cheb Basin (Czech Republic). J. Geodyn. 35:5–32

Barsch D (1969) Studien zur Geomorphogenese des zentralen Berner Juras. Basler Beitr. Geogr. 3:1–221

Bell JS, Gough DI (1979) Northeast-southwest compressive stress in Alberta: evidence from oil wells. Earth Planet. Sci. Lett. 45:475–582

v. Bertalanffy L (1932) Theoretische Biologie. Springer, Berlin

Boegel H, Schmidt K (1976) Kleine Geologie der Ostalpen. Ott Verlag, Thun

Bold HC (1966) Mauritius – Physical Geography. Encyclop. Brit. (Univ. Chicago) 14:1129–1130

Bonatti E (1996) Long-lived oceanic transform boundaries formed above mantle thermal minima. Geology 24(9):803–806

Bondam J (1975) Greenland. In: Fairbridge RW (ed) Encyclopedia of World Regional Geology. Dowden, Hutchinson and Ross, Stroudsburg, Pa., 1, pp 292–300

Bonneton JR, Scheidegger AE (1981) Relations between fracture patterns, seismicity and plate motions in the Lesser Antilles. J. Struct. Geol. 3(4):359–369

Bonneton JR, Scheidegger AE (1982) La fracturation de type "diaclase" dans l'archipel de la Guadeloupe: implications géotectoniques. Geol. Appl. Idrogeol. Bari 17:203–241

Braithwaite CJR (1984) Geology of the Seychelles. In: Stoddard DR (ed) Biogeography and ecology of the Seychelles Islands. Junk, The Hague, pp 17–38

Briceno HO, Schubert C (1985) Analisis del fracturamiento en zonas de Tepuí, Estado Bolivar, Venezuela. Mem. VI Congreso Geológico Venezolano, Caracas 7, 5603–5621

Brix F (1970) Der Raum von Wien im Laufe der Erdgeschichte. In: Starmühlner F, Ehrendorfer F (eds) Naturgeschichte Wiens, Band 1. Heimat und Volk, Wien, Muenchen, pp 27–234

Brix F (1972) Naturgeschichte Wiens, Band III, Kartenteil. Jugend und Volk, Wien

Brun E (1986) Ordnungs-Hierarchien. Neujahrsblatt. Naturforsch. Gesellsch. Zürich, 188. Stück Orell-Fuessli, Zuerich

Buser H (1966) Paleostructures of Nigeria and Adjacent Countries. In: Geotekt. Forsch. (Stille H, Lotze F (eds); Publ. Schweizerbart, Stuttgart), vol 24: 1–90

Buser H, Scheidegger AE (1992) Mechanik einer Rutschung zwischen Sattel und Schwyz (Kanton Schwyz, Schweiz). Interpraevent 5:169–178

Byerly P (1938) The earthquake of July 6, 1934: Amplitudes and first motion. Bull. Seismolog. Soc. Am. 28:1–13

Çambel AB (1993) Applied Chaos Theory. Boston, Academic Press

Caputo R (1995) Evolution of orthogonal sets of coeval extension joints. Terra Nova 7(5):479–490

Carniel P, Hauswirth EK, Roch KH, Scheidegger AE (1975) Geomechanische Untersuchungen in einem Rutschungsgebiet im Felbertal in Oesterreich. Verh. d. Geol. Bundesanst. 1975(4):305–330

Carson MA, Kirkby MJ (1972) Hillslope form and process. University Press, Cambridge

Castillo E (1997) Untersuchung von Massenbewegungen mit geophysikalischen Methoden und FE-Modellrechnungen. PhD Thesis, Section of Geophysics, Tech. Univ. Vienna

Chevalier JP (1975a) Tahiti. In: Fairbridge RW (ed) Encyclopedia of World Regional Geology. Dowden, Hutchinson and Ross, Stroudsburg, Pa., 1, p 493

Chevalier JP (1975b) Society Islands. In: Fairbridge RW (ed) Encyclopedia of World Regional Geology. Dowden, Hutchinson and Ross, Stroudsburg, Pa., 1, pp 449–450

Christie RL (1975) Canada-Arctic Archipelago. In: Fairbridge RW (ed) Encyclopedia of World Regional Geology. Dowden, Hutchinson and Ross, Stroudsburg, Pa., 1, pp 145–156

Chronic H (1980) Roadside Geology of Colorado. Mountain Press, Missoula Mont.

Chubb LJ (1934) The structure of the Pacific Basin. Geol. Mag. 64:518–522

Cotton CA (1944) Volcanic Landscape Forms, Whitcombe and Tombs, Christchurch

Cotton CA (1968) Tectonic landscapes. In: Fairbridge RW (ed) Encyclopedia of Geomorphology. Reinhold, New York, pp 1109–1117

Cramer F (1993) Chaos and Order. Translated into German by Loewus DIF. VCH, Weinheim

Cumberland KB (1966) New Zealand, Structure, Geology, and Landforms. Encyclopedia Britannica 16:449–450

D'Addario GW (1975) Australia-Northern Territory. In: Fairbridge RW (ed) Encyclopedia of World Regional Geology. Dowden, Hutchinson and Ross, Stroudsburg, Pa., 1, pp 46–56

Dalziel IWD (ed) (1989) Tectonics of the Scotia Arc, Antarctica. Field Trip Guidebook T180, Am. Geophys. Union, Washington

Davies TRH (1985) Mechanics of large debris flows. In: Takei A (ed) Proc. Internat. Symposium on Erosion, Debris Flow and Disaster Prevention, Sept. 3–5, 1985, Tsukuba, Japan, Toshindo Printers, Tokyo, pp 215–218

Davis WM (1909) Geographical Essays. Ginn, Boston (Reprint Dover, New York, 1954)

Davis WM (1924) Die erklärende Beschreibung der Landformen, 2.Aufl. Teubner, Leipzig

Decker R, Decker B (1986) Hawaii Volcanoes National Park and Vicinity, Hawaii, 1:100000-scale topographic map. Denver, U.S. Geol. Survey, Denver.

DeGraff JM, Aydin A (1987) Surface morphology of columnar joints and its significance to mechanics and direction of joint growth. Bull. Geol. Soc. Am. 99:605–617

Denham D, Alexander LG, Worotnicki G (1979) Stresses in the Australian Crust, evidence from earthquakes and in-situ stress measurements. BMR J. Aust. Geol. Geophys. 4:189–195

De Villiers AB, Scheidegger AE, Beckedahl HR (1993) Neotectonic joint and drainage patterns in the Vredefort Dome structure, South Africa. Geoökodynamik 14:105–114

Dickinson WR, Hopson CA, Saleeby JB (1996) Alternate origins of the Coast Range Ophiolite (California): Introduction and implications. GSA Today 6(2):1–10

Dietz R (1961) Ocean basin evolution by spreading of the sea floor. Nature 190:854–857

Dietz RS (1961) The Vredefort Rong Structure: A meteorite impact scar? J. Geology 69, 499–516

Dikau R, Brunsden D, Schrott L, Ibsen ML (eds) (1996) Landslide Recognition: Identification, Movement and Causes. Wiley, New York

Dimroth E (1963) Fortschritte der Gefügestatistik. Neues Jahrbuch der Mineralogie Monatshefte 1963:186–192

Druitt H, Sparks RSJ (1984) On the formation of calderas during ignimbrite eruptions, Nature: 310 (5979):679–681

Du Toit AL (1939) The Geology of South Africa, 2nd ed. Oliver and Boyd, Edinburgh

Dutta P, Horn PM (1981) Low-frequency fluctuations in solids: 1/f noise. Revs. Mod. Phys. 53:497–516

Eberhardt M (1999) Der Felssturz/Erdrutsch Hinter Königstein in Küttigen. Mitt. Aargauischen Naturforsch. Gesellsch. 35:133–146

Engelder T (1985) Loading paths to joint propagation during a tectonic cycle: an example from the Appalachian Plateau, U.S.A. J. Struct. Geol. 7:459–476

Erlach R, Scheidegger AE (1983) Joint orientations in the Southern Sahara. Arch. Met. Geoph. Biocl. Ser.A 32:191–196

Evison FF, Webber SJ (1986) Seismogenic stress in Central New Zealand. Bull. Roy. Soc. New Zealand 24:553–565

Eyles N, Arnaud E, Scheidegger AE, Eyles C (1997) Bedrock jointing and geomorphology in southwestern Ontario, Canada: an example of tectonic predesign. Geomorphology 19:17–34

Eyles N, Clark BM, Kaye BG, Howard KWF, Eyles CH (1985a) The application of basin analysis techniques to glaciated terrains: An example from the Lake Ontario Basin. Geosci. Canada 12(1):22–32

Eyles CH, Eyles N (1983) Sedimentation in a large lake: A reinterpretation of the late Pleistocene stratigraphy at Scarborough Bluffs, Ontario, Canada. Geology 11:146–152

Eyles N, Eyles CH, Lau K, Clark B (1985b) Applied sedimentology in an urban environment – the case of Scarborough Bluffs, Ontario; Canada's most intractable erosion problem. Geosci. Canada 12(3):91–104

Eyles N, Scheidegger AE (1995) Environmental significance of bedrock jointing in Southern Ontario, Canada. Environ. Geol. 26:269–277

Eyles N, Scheidegger AE (1999) Neotectonic jointing control on Lake Ontario shoreline orientation at Scarborough, Ontario. Geosci. Canada 26(1):27–30

Fairbridge RW (1975) Oceania. In: Fairbridge RW (ed) Encyclopedia of World Regional Geology. Dowden, Hutchinson and Ross, Stroudsburg, Pa., 1, pp 410–415

Fellenius W (1927) Erdstatische Berechnungen. Ernst, Berlin

Fernandez JC (Ed.) (1981) Geology and Mineral Resources of the Philippines. Vol. 1. Bureau of Mines, Manila

Figdor H, Hauswirth EK, Lindner H, Roch KH, Scheidegger AE (1990) Geodaetische und geophysikalische Untersuchungen am NW-Hang des Graukogels bei Badgastein. Oesterr. Z. für Vermessungswesen & Photogrammmerie: 78(2):59–76

Figdor H, Lahodynsky R, Roch KH, Scheidegger AE (1983) Untersuchung eines Luftbild-lineamentes am Sandberg, Verw. Bez. Hollabrunn. Unsere Heimat (Wien) 54(2):109–119

Fränzle O (1968) Valley evolution. In: Fairbridge RW (ed) Encyclopedia of Geomorphology. Reinhold Book Co., New York, pp 1183–1189

Gees RA (1969) The age of the Bermuda seamount. Maritime Sed. 5(2):56–57

Gees RA, Medioli F (1970) A continuous survey of the Bermuda Platform, Parts I and II. Marine Sedimentation 6(1):21–25 and 6(3):118–120

Geotimes (1980) Geologic events reported by the Scientific Alert Network NHB9, Smithsonian Inst. Washington. Geotimes 25(3), 22

Gerber E (1945) Lage und Gliederung des Lauterbrunnentales und seiner Fortsetzung bis zum Brienzersee. Mitt. Aargauischen Naturf. Gesellsch. 22:165–184

Gerber E (1963) Ueber Bildung und Zerfall von Waenden. Geogr. Helv., 18, 331–35

Gerber E (1969) Bildung von Gratgipfeln und Felswänden in den Alpen. Geomorph. Supplementband 8:94–118

Gerber EK, Scheidegger AE (1974) On the dynamics of scree slopes. Rock Mech. 6, 25–38

Gerber E, Scheidegger AE (1975) Geomorphological evidence for the geophysical stress field in mountain massifs. Riv. Ital. Geofis. Sci Affini, 2(1):47–52

Gerber EK, Scheidegger AE (1984) Eine chronische Rutschung in tonigem Material. Vierteljahrsschr. Naturf. Gesellsch. Zürich 129(3):294–315

Ghose G, Yoshioka, Oike,K. (1990) Three-dimensional numerical simulation of the subduction dynamics in the Sunda Area region, Southeast Asia. Tectonophysics 181:223–255

Gowd TN, Rao MV, Krishnamurthy R, Rummel F, Alheid HJ (1981) In-Situ Stress Measurements by Hydraulic Fracturing in the Underground Mines of Kolar Gold Field, India. Rept. to Indo-German Collaboration Project

Gregory JW (1913) The Nature and Origin of Fjords. Murray, London

Guille G, Goutière G, Sornein JF (1993) Les atolls de Mururoa et de Fangataufa: Géologie-Pétrologie-Hydrogéologie. Editions du Commissariat à l'Energie Atomique/Direction Centrale, Brouillère-le-Châtel

Gubler E (1976) Beitrag des Landesnivellements zur Bestimmung vertikaler Krustenbewegungen in der Gotthard-Region. Schweiz. mineral. petrogr. Mitt. 56:675–678

Haken H, Wunderlin A (1991) Die Selbstrukturierung der Materie. Vieweg, Braunschweig

Hancock PL (1991) Determining contemporary stress directions from neotectonic joint systems. Phil. Trans. R. Soc. Lond. 337:29–40

Hancock PL, Engelder T (1989) Neotectonic joints. Bull. Geol. Soc. Am. 101:197–208

Hantke R (1978) Eiszeitalter: Die jüngste Erdgeschichte der Schweiz und ihrer Nachbargebiete, Vol. 1. Ott, Thun (reprint 1993 by Ecomed, Landsberg/Lech)

Hantke R (1991) Landschaftsgeschichte der Schweiz und ihrer Nachbargebiete. Ott, Thun

Hantke R (1993) Flussgeschichte Mitteleuropas. Enke, Stuttgart

Hantke R, et al. (1967) Geologische Karte des Kantons Zürich und seiner Nachbargebiete 1:50000 in 2 Blättern mit Erläuterungen. Vierteljahrsschr. natf. Gesellsch. Zürich 112(2)

Hantke R, Mueller ER, Scheidegger AE, Wiesmann A (2003) Der Molasse-Schuttfächer des Otteberge und der Lauf der Thur seit dem jüngeren Tertiär. Mitt. der Thurgauischen Naturf. Gesellsch. 59:85–111

Hantke R, Scheidegger AE (1993) Zur Genese der Aareschlucht. Geogr. Helv. 48/3:120–124

Hantke R, Scheidegger AE (1994) Klusen und verwandte Formen im Schweizer Jura. Geogr. Helv. 49(4):157–164

Hantke R, Scheidegger AE (1997) Zur Morphogenese der Zürichseetalung. Vierteljahrsschr. Naturf. Gesellsch. Zuerich 142(3):89–95

Hantke, R, Scheidegger AE (1998) Morphotectonics of the Mascarene Islands. Annali di Geofis. (Roma) 41(2):165–181

Hantke R, Scheidegger AE (1999) Tectonic predesign in geomorphology. In: Hergarten S, Neugebauer HJ (eds) Process Modelling and Landform Evolution, Lecture Notes in Earth Sciences, no. 78. Springer Verlag, Berlin-Heidelberg-New York, pp. 252–266

Hantke R, Scheidegger AE (2001) Zur Entstehung der Taminaschlucht. Terra Plana 2001(1), 30–34

Hantke R, Scheidegger AE (2003) Zur Morphotektonik der Zentralschweizerischen Alpenrandseen. Berichte der Schwyzerischen Natf. Gesellsch. 14:83–98

Harrison RG, Biswas DJ (1986) Chaos in light. Nature 321:394–401

Harvey D (1969) Explanation in geography. Arnold, London

Hauswirth EK, Lahodynsky R, Roch KH, Scheidegger AE (1982) Geophysikalische Untersuchungen an der Grosshangbewegung Wörschachwald (Ennstal, Stmk.). Mitt. Naturwiss. Ver. Steiermark 112:75–90

Hauswirth EK, Pirkl H, Roch KH, Scheidegger AE (1979) Untersuchungen eines Talzuschubes bei Lesach (Kals, Osttirol). Verh. Geol. Bundes-Anst. Wien 1979(2):51–76

Hauswirth EK, Scheidegger AE (1980) Tektonische Vorzeichnung von Hangbewegungen im Raume von Badgastein. INTERPRAEVENT 1:159–178

Hast N (1958) The Measurement of Rock Pressure in Mines. Sveriges Geologiska Undersok., Stockholm

Hausdorff F (1919) Dimension und äusseres Mass. Math. Ann. 79:157–179

Hawkins JW, Natland JH (1975) Nephelinites and basanites of the Samoan linear volcanic chain: their possible tectonic significance. Earth Planet. Sci. Lett. 24:427–439

Heim A (1893) Die Entstehung der alpinen Randseen. Vierteljahrsschr. natf. Ges. Zürich 39/1:65–84

Heirtzler JR, Dixon GO, Herron EM, Pitman, WC III, LePichon X (1968) Marine magnetic anomalies, geomagnetic reversals, and motions of the ocean floor and continents. J. Geophys. Res. 73:2119–2136

Hess HH (1962) History of ocean basins. In: Engel AEJ, James HI, Leonard BF (eds) Petrologic Studies: a volume in honor of A. F. Buddington. Geol. Soc. America 599–620

Heuberger H (1975) Das Oetztal. Bergstürze und Gletscherstände, kulturgeographische Gliederung. Innsbrucker Geogr. Studien 2:213–249

Hielle A (1993) Geology of Svalbard. Norsk Polarinstitutt, Oslo

Higgins M, Higgins R (1996) A Gelogical Companion to Greece and the Aegean. Duckworth, London

Hobbs WH (1912) Earth Features and their Meaning. Macmillan, New York

Holmes A (1944) Principles of Physical Geology, 1st Edition. Nelson, London

Holmes A (1965) Principles of Physical Geology, 2nd, Completely Revised Edition. Nelson, London

Holtedahl H (1967) Notes on the formation of fjords and fjord-valleys. Geogr. Ann. 49A(1–2):188–203

Horton RE (1945) Erosional development of streams and their drainage basins; hydrophysical approach to quantitative morphology. Bull. Geol. Soc. Am. 56:275–370

Hough JL (1968) Great Lakes (North America). In: Fairbridge RW (ed) Encyclopedia of Geomorphology. Reinhold, New York, pp 499–506

Hutter K, Vulliet L (1985) Gravity-driven slow creeping flow of a thermoviscous body at elevated temperatures. J. Therm. Stresses 8:99–138

Jacobshagen V (Ed.) mit Beiträgen von U. Dornsiepen (1986) Geologie von Griechenland. Borntraeger, Berlin

Jacobshagen V (1986) Kreta. In: Geologie von Griechenland. Borntraeger, Berlin 55–80

Johnson MD, Armstrong DK, Sanford BV, Telford PG, Rutka MA (1992) Paleozoic and Mesozoic Geology of Ontario. In: Geology of Ontario. Ontario Geol. Survey, Special Vol. 4, Part 2, pp 907–1010

Jonsson S, Alves MM, Sigmundsson F (1999) Low rates of deformation of the Furnas and Fogo volcanoes, Sao Miguel, Azores, observed with the Global Positioning System 1993–1997. J. Volcanol. Geotherm. Res. 92:83–94

Kauffman SA (1993) The Origins of Order: Self-Organization and Selection in Evolution. Oxford University Press, Oxford

Kear D, Wood BL (1959) The Geology and Hydrology of Western Samoa. New Zeal. Geol. Surv. Bull. n.s.63; Wellington

Kebede F (1989) Seismotectonic studies of the East African Rift system north of 12°S to the southern Red Sea. Report No. 1–89, Seismological Department, Uppsala

Kent LE (1980) Stratigraphy of South Africa. Department of Mineral and Energy Affairs, Geol. Surv., Handbook 8, Govt. Printer, Pretoria

King LC (1966) Africa, Geology. Encyclop. Brit. 1:147–249

Kizaki K (1994) An Outline of the Himalayan Upheaval. Japan International Cooperation Agency, Kathmandu

Kogan M, Steblov GM, King RW, Herring TA, Frolov DI, Egorov SG, Levin YI, Lerner-Lam A, Jones A (2000) Geodetic constraints on the rigidity and relative motion of Eurasia and America. Geophy. Res. Lett. 27(14):2041–2044

Kohlbeck FK, Mojica J, Scheidegger AE (1994) Clast orientations of the 1985 lahars of the Nevado del Ruiz, Colombia and implication for depositional processes. Sedimentary Geol. 88:175–183

Kohlbeck FK, Scheidegger AE (1977) On the theory of the evaluation of joint-orientation measurements. Rock Mechanics 9:9–25

Kohlbeck FK, Scheidegger AE (1985) The power of parametric orientation statistics in the Earth Sciences. Mitt. Österr. geol. Gesellsch. Wien 11:251–265

Kopp J (1962) Veränderungen von Seen und Flussläufen in der Zentalschweiz in interglazialer und postglazialer Zeit. Mitt. Naturf. Gesellsch. Luzern 19:153–166

Labhart TP (1991) Geologie der Schweiz. Ott, Thun

Land LS, Mackenzie FT (1970) Field Guide to Bermuda Geology. Bermuda Biological Station Spec. Publications, Hamilton, #4

Langbein WB, Leopold LB (1966) River meanders, theory of minimum variance. U.S. Geol. Surv. Prof. Pap. 422-H

Langel RA, Thorning L (1982) A satellite magnetic anomaly map of Greenland. Geophys. J. Roy. Astr. Soc. 71:599–612

Lattman L (1968) Structural control in Geomorphology. In: Fairbridge RW (ed) Encyclopedia of World Regional Geomorphology. Reinhold Book Co., New York, pp 1074–1079

Laubscher H (1987) Die tektonische Entwicklung der Nordschweiz. Eclogae Geol. Helv. 80(2), 287–303

Le Bas MJ, Rex DC, Stillman CJ (1986) The early magmatic chronology of Fuerteventura, Canary Islands. Geol. Mag. London 123(3):287–298

Lechthaler-Zdenkovic M, Scheidegger AE (1989) Entropy of landscapes. Z. für Geomorphologie 33(3):361–371

Legrand P (1982) Essai sur la paléogéographie du silurien au Sahara algérien. Not. Mem. Comp. Franc. Petroles (Paris) 16:9–24

Leopold LB, Langbein WB (1962) The concept of entropy in landscape evolution. U.S. Geol. Surv. Prof. Pap. 500A:A1–A20

Liao KH, Scheidegger AE (1968) A computer model for some branching-type phenomena in hydrology. Bull. Int. Assoc. Sci. Hydrol. 13(1):5–13

Louis H (1979) Allgemeine Geomorphologie. Walter de Gruyter, Berlin

Lorenz EN (1963) Deterministic nonperiodic flow. J. Atmos. Sci. 357:130–141

Lundqvist S, March NH, Tosi M (eds) (1988) Order and Chaos in Nonlinear Physical Systems. Plenum Press, New York/London

Ma ZJ, Zhang, P.Zh., Hong HJ, Gao XL (1997) Some fundamental problems of geodynamics and application of Earth observations by space-based methods. Earthquake Research China 11(1):1–11

Machatschek F (1952) Geomorphologie, 5. Aufl. Teubner, Leipzig

Magilligan FJ (1992) Thresholds and spacial variability of flood power during extreme floods. Geomorphology 5:373–390

Maiti GS (1980) Quantitative analysis of the Jaldhaka Basin. Indian J. Landscape Systems and Ecological Studies 3(1–2):58–66

Maiti GS (1991) An Analytical Study of the Landforms and Land Uses of the Tarai Area (The Eastern Himalayas), West Bengal. PhD Thesis, Dept. Geography, Calcutta Univ., Calcutta

Mandelbrot BB (1967) How long is the coast of Britain? Statistical self-similarity and fractonal dimension. Science 155:636–638

Marsal D (1967) Statistische Methoden für Erdwissenschaftler. Schweizerbart, Stuttgart

Marshall P (1916) Oceania. Handbuch Reg.Geol. 7(9)

Martinez de Pison E, Quirantes F (1994) Relieve de las islas Canarias. In: Gutierrez-Elorza M (ed) Geomorfologia de Espana. Ediciones Rueda, Madrid, pp 495–526

Mayer H, Duarte J, Paraffan A (1986) Caracteristicas físicas del sismo de Popayán. In: El sismo de Popayán del 31 de Marzo de 1983. Ingeominas, Bogotá, pp 119–147

McBirney AR, Murase T (1984) Rheological properties of magmas. Annu. Rev. Earth Planet. Sci. 1984:337–358

McTigue DF, Mei CC (1981) Gravity-induced stress near topography of a small slope. J. Geophys. Res. 86(B10): 9268–9278

Meco J, Stearns CE (1981) Emergent littoral deposits in the Eastern Canary Islands. Quaternary Res. 15:199–208

Medwenitsch W (1970) Zur Geologie und regionalen Stellung der Canarischen Inseln. Mitt. Geol. Gesellsch. Wien 63:160–184

Meyerhoff AA, Agocs WB, Taner I, Morris AEL, Martin BC (1992a) Origin of midocean ridges. In: Chatterjee S, Hotton N (eds) New Concepts in Global Tectonics. Texas Tech. Univ. Press, Lubbock, pp 151–178

Meyerhoff AA, Taner I, Morris AEL, Martin BD, Agocs WB, Meyerhoff HA (1992b) In: Chatterjee S, Hotton N (eds) New Concepts in Global Tectonics. Texas Tech. Univ. Press, Lubbock, pp 309–409

Mezcua J, Buforn E, Udias A, Rueda J (1992) Seismotectonics of the Canary Islands. Tectonophysics 208:447–452

Milne G (1935) Some suggested units of classification and mapping, particularly for East African soils. Soil Res. 4:183–98

Milne G (1947) A soil reconnaisance journey through parts of Tanganyika Territory. J. Ecol. 27:192–265

Mitchell-Thomé RC (1976) Geology of the Middle Atlantic Islands. Borntraeger, Berlin

Mohr O (1928) Abhandlungen aus dem Gebiete der technischen Mechanik, 3.Aufl. Wilh. Ernst and Sohn, Berlin

Mohr P (1983) Ethiopian flood basalt province. Nature 303:577–584

Mohr P, Mitchell JG, Raynolds RGH (1980) Quaternary volcanism and faulting at O'a caldera, Central Ethiopian Rift. Bull. Volcanol. 43(1):173–189

Mojica J, Scheidegger AE (1981) Diaclasas recientes en Colombia y su siginificado tectónico. Geologia Colombiana 12:75–90

Monin AS (1991) Union Lecture: Predictability of chaotic phenomena. Chronique de l'Union Géodés. Géophys. Internat. 208:268–282

Mountjoy AB (1966) Egypt, Geology and Structure. Encyclop. Brit. 8:27–28

Mukhopadhyay SC (1982) The Tista Basin, a Study in Geomorphology. Bagchi & Co., Calcutta

Müller F (1938) Das Gebiet der Aareschlucht. p. 42–47 in Geologie der Engelhörner, der Aareschlucht und der Kalkkeile bei Innertkirchen (Berner Oberland). Beiträge zur Geol. Karte der Schweiz N.F. 74, I-X, 1–55

Müller-Salzburg L (1963) Der Felsbau. F. Enke, Stuttgart

Nicolaysen LO, Ferguson J (1981) Diapirs driven by high fluid pressure. J. Struct. Geol. 3:89–95

Nicolis G, Prigogine I (1977) Self-Organization in Non-Equilibrium Systems, 8th printing. Wiley, New York

Norris RM, Webb RW (1990) Geology of California. Wiley, New York

Noverraz F, Bonnard C (1990) Technical note on the visit of the La Frasse Landslide. In: Bonnard C (ed) Proc. 5th Internat. Sympos. Landslides Lausanne 10–15 July 1988. Balkema, Rotterdam, 3:1549–1554

Nunn P (1988) Vatulele: A study in the geomorphological development of a Fiji island. Memoir 2, Mineral Resources Dept., Suva

Nunn P (1994) Oceanic Islands. Blackwell, Oxford

Nye JF (1952) The mechanics of glacier flow. J. Glaciol. 2(12):82–93

Okagbue CO (1989) Predicting landslides caused by rainstorms in residual-colluvial soils of Nigerian hillside slopes. Natural Hazards 2:133–141

Okal E, Talandier J, Sverdrup KA, Jordan TH (1980) Seismicity and tectonic stress in the South-Central Pacific. J. Geophys. Res. 5(B11):6479–6495

Ollier C (1981) Tectonics and Landforms. Longman, London

Orvin AK (1966) Spitsbergen, Physical features; structure and geology. Encyclop. Brit., Chicago 21, 242

Paquin C, Bloyet J, Angelidis C (1984) Tectonic stresses on the boundary of the Aegean domain: in situ measurements by overcoring. Tectonophysics 110:145–150

Park S, Iversen JD (1984) Dynamics of lava flow: thickness growth characteristics of steady two-dimensional flow. Geophys. Res. Lett. 11(7):641–644

Penck A, Brueckner E (1909) Die Alpen im Eiszeitalter. Tauchnitz, Leipzig

Pennington WD (1981) Subduction of the Eastern Panama Basin of northwestern South America. J. Geophys. Res. 86:10753–10770

Pflueger F, Seilacher A (1991) Flash flood conglomerates. In: Einsele G, Ricken W, Seilacher A (eds) Cycles and Events in Stratigraphy. Springer, Berlin, pp 383–391

Pinsker LM (2003) Bending thoughts of the Hawaiian Chain. Geotimes 48(3): 29&46

Pizzuto J (1992) The morphology of graded rivers: a network perspective. Geomorphology 5:457–474

Ploechinger B, Prey S (1974) Der Wienerwald. Sammlung Geologischer Führer, Bd. 59. Borntraeger Berlin/Stuttgart

Poldervaart A (1971) Volcanicity and forms of extrusive bodies. In: Green J, Short NM (eds): Volcanic Landforms and Surface Features. Springer, Berlin, New York, p. 1–18

Pollard DD, Aydin A (1988) Progress in understanding jointing over the past century. Bull. Geol. Soc. Am. 100:1181–1204

Popper KR (1984) Logik der Forschung, 8th ed. Mohr, Tübingen

Prasad N (1979) Jointing controls on drainage in the Barakar Basin. Geogr. Outlook (Ranchi, India) 14:101–106

Price NJ (1966) Fault and joint development in brittle and semi-brittle rock. Pergamon Press, Oxford

Prigogine I (1947) Étude thermodynamique des phénomènes irreversibles. Desoer, Liège

Ranalli G, Scheidegger AE (1968) A test of the topological structure of river nets. Bull. Int. Assoc. Sci. Hydrol. 13(2):142–153

Rehbock R (1929) Bettbildung, Abfluss und Geschiebebewegungen bei Wasserläufen. Z. Deutsch. Geol. Ges. 1929:498–534

Reiter M, Barroll MW, Minier J, Clarkson G (1987) Thermo-mechanical model for incremental fracturing in cooling lava flows. Tectonophysics 142:241–260

Richards HG (1975) Easter Island and Sala y Gomez. In: Fairbridge RW (ed) Encyclopedia of World Regional Geology. Dowden, Hutchinson & Ross, Stroudsburg, Pa., 1, pp 260–261

Rickard MJ (1975) Australasia – regional review. In: Fairbridge RW (ed) Encyclopedia of World Regional Geology. Dowden, Hutchinson & Ross, Stroudsburg, Pa., 1, pp 21–28

Rodda P (1994) Geology of Fiji. SOPAC Techn. Bull. 8:131–151

Rogojina C (1993) Neotectonic Bedrock-Joints and Pop-Ups in the Metropolitan Toronto Area. Geology Dept.,Toronto

Rothe P (1996) Kanarische Inseln. Sammlung Geologischer Führer, Bd. 81, 2nd ed. Born-
    traeger, Berlin/Stuttgart
Rust BR (1972) Pebble orientations in fluvial sediments. J. Sediment. Petrol. 42(2):384–388
Sah MP, Bist KS (1998) Field Guide on Okimath Area, Garhwal Himalaya. Wadia Institute of
    Himalayan Geology, Dehra Dun
Said R (1962) The Geology of Egypt. Elsevier, Amsterdam
Scarth A, Tanguy JC (2001) Volcanoes of Europe. University Press, Oxford
Schaefer KH (1978) Geodynamik an Europa's Plattengrenzen. Fredericiana 23:30–46
Scheidegger AE (1961) Mathematical models for slope development. Bull. Geol. Soc. Am.
    72:37–50
Scheidegger AE (1965) On the statistics of the orientation of bedding planes, grain axes,
    and similar sedimentological data. U.S. Geol. Surv. Prof. Pap. 525C:C164–C167
Scheidegger AE (1967a) A complete thermodynamic analogy for landscape evolution. Bull.
    Int. Assoc. Sci. Hydrol. 12(4):57–62
Scheidegger AE (1967b) A stochastic model for drainage patterns into an intramontane
    trench. Bull. Int. Assoc. Sci. Hydrol. 12(1):15–20
Scheidegger AE (1977) Joints in the Bahamas and their geotectonic significance. Riv. Ital.
    Geofis. Sci. Affini 4(3/4):109–118
Scheidegger AE (1978) The enigma of jointing. Riv. Ital. Geofis. Sci. Affini 5:1–4
Scheidegger AE (1979a) Beziehungen zwischen Orientationsstruktur der Talanlagen und
    der Kluftstellungen in Österreich. Mitt. Österr. Geogr. Gesellsch. 121(2):187–195
Scheidegger AE (1979b) Orientationsstruktur der Talanlagen in der Schweiz. Geogr. Helv.
    34(1):9–15
Scheidegger AE (1979c) Joints and valley trends in Spain. Geol. Appl. Idrogeolog.
    14(1):167–179
Scheidegger AE (1979d) On the tectonics of the Western Himalaya. Arch. Met. Geophys.
    Biokl. A28:89–106
Scheidegger AE (1979e) The principle of antagonism in the Earth's evolution. Tectono-
    physics, 55:T7–T10
Scheidegger AE (1980) Joint orientation measurements in Australia. Geol. Appl. Idrogeolog.
    (Bari) 15:121–146
Scheidegger AE (1981) Joint orientation measurements in Western Alberta and British
    Columbia. Ann. Geofis. Roma 34:55–62
Scheidegger AE (1982a) Principles of Geodynamics, 3rd edition. Springer, Berlin, New York
Scheidegger AE (1982b) Joint-orientations in Ethiopia. Arch. Meteorol. Geophys. Biokl.
    Ser.A 31:269–272
Scheidegger AE (1983a) Instability principle in geomorphic equilibrium. Z. für Geomor-
    phologie 27(1):1–19
Scheidegger AE (1983b) Interpretation of fracture and physiographic patterns in Alberta,
    Canada. J. Struct. Geol. 5(1):53–59
Scheidegger AE (1985a) Bergsturz Amden 1974. Geogr. Helv. 40(1):25–29
Scheidegger AE (1985b) Garlands of heads. Z. für Geomorphologie 29(2):223–234
Scheidegger AE (1985c) The significance of surface joints. Geophys. Surv. 70:259–271 (1.42)
Scheidegger AE (1986) The catena principle in geomorphology. Z. für Geomorphologie
    30(3):257–273
Scheidegger AE (1987) The fundamental principles of landscape evolution. Catena Suppl.
    10:199–210
Scheidegger AE (1989) Estudios tectónicos de la parte sur de México. Rev. Acad. Colomb.
    Cienc. Exactas, Fis. Nat. 17(64):125–132

Scheidegger AE (1990) Joints and the neotectonics of the Scotia Plate, Antarctica. In: Rossmanith HP (ed) Mechanics of Jointed and Faulted Rocks. Balkema, Rotterdam, pp 187–193

Scheidegger AE (1991) Theoretical Geomorphology, 3rd completely revised edition. Springer, Berlin, New York

Scheidegger AE (1993a) Joints as Neotectonic Signatures. Tectonophysics 219:235–239

Scheidegger AE (1993b) On the genesis of lava cracks in Hawaii. Terra Nova 5:560–562

Scheidegger AE (1995) Geojoints and geostresses. In: Rossmanith HP (ed) Proc. 2nd Intern. Conf. Mechanics of Jointed and Faulted Rocks. Balkema, Rotterdam, pp 3–35

Scheidegger AE (1996) Ordnung am Rande des Chaos, ein neues Naturgesetz. Österr. Z. für Vermessungswesen und Geoinformation 84(1):89–74

Scheidegger AE (1998a) Morphotectonic indications for the opening of Davis Strait. In: Rossmanith HP (ed) Proc. 3rd Intern. Conf. Mechanics of Jointed and Faulted Rocks. Balkema, Rotterdam, pp 95–100

Scheidegger AE (1998b) Morphotektonik am Westrand des Wiener Beckens. Österr. Z. für Vermessung und Geoinformation 86(2):92–100

Scheidegger AE (1999a) Note on the morphotectonics of the Kaliasaur slide, Garhwal Himalaya, Uttar Pradesh, India. Himalayan Geol. 20(2):105–107

Scheidegger AE (1999b) Morphotectonics of Eastern Nepal. Indian J. of Landscape Systems and Ecol. Stud. 22(2):1–9

Scheidegger AE (2000a) Tektonische Aspekte von Bergstürzen. INTERPRAEVENT 1:339–347

Scheidegger AE (2000b) Morphotectonics of mass movements on slopes. J. Nepal Geol. Soc. 22:365–370

Scheidegger AE (2001a) Surface joint systems, tectonic stresses and geomorphology: a reconciliation of conflicting observations. Geomorphology 38(3–4): 213–219

Scheidegger AE (2001b) Morphotectonics of the Seychelles Islands in the Indian Ocean. Indian J. of Landscape Devolopment and Ecol. Stud. 24(2):1–9

Scheidegger AE (2002a) "Inselberge" im Umfeld von Wien. Österr. Z. für Vermessung und Geoinformation 90(1):13–22

Scheidegger AE (2002b) Morphometric analysis and its relation to tectonics in Macaronesia. Geomorphology 46(1–2): 95–115

Scheidegger AE, Ai NS (1986) Tectonic processes and geomorphological design. Tectonophysics 126:285–300

Scheidegger AE, Ai NS (1987) Clay slides and debris-flows in the Wudu Region and their tectonic implications. Science Exploration (China) 7(1):253–264

Scheidegger AE, Ajakaiye DE (1985) Geodynamics of Nigerian shield areas. J. Afr. Earth Sci. 3(4):461–470

Scheidegger AE, Ajakaiye DE (1990) Mass movements in flat sedimentary savannah areas. Indian J. Landscape Systems and Ecol. Stud. 13(1):1–9

Scheidegger AE, Ajakaiye DE (1994) Mass movements in hilly areas (with examples from Nigeria). Natural Hazards 9:191–196

Scheidegger AE, Hantke R (1994) On the genesis of river gorges. Trans. Japanese Geomorphol. Union, 15(2):91–110

Scheidegger AE, Hauswirth EK, Lahodynsky R (1984) Geowissenschaftliche Untersuchungen als Grundlage für Schutzmassnahmen im Bereich des Mödritschbaches. INTERPRAEVENT 2:9–20

Scheidegger AE, Kohlbeck FK (1985) The selection pronciple in surface erosion. In: Takei A (ed) Proc. Internat. Symp. on Erosion, Debris Flow and Disaster Prevention, Tsukuba, Japan. Toshindo Printers, Tokyo, pp 285–290

Scheidegger AE, Padale JG (1982) A geodynamic study of Peninsular India. Rock Mech. 15:209–241

Scheidegger AE, Schubert C (1989) Neotectonic provinces and joint orientations of northern South America. J. South Amer. Earth Sci. 2(4):331–341

Scheidegger AE, Turek A (1978) Joints in Eastern Manitoba. Arch. Met. Geoph. Biokl. A27:381–389

Schuster HG (1984) Deterministic Chaos. Physik, Berlin

Schluechter C (1983) Die Bedeutung der angewandten Quartärgeologie für die eiszeitgeologische Forschung in der Schweiz. Phys. Geogr. (Zürich) 11:59–72

Schluechter C (1987) Talgenese im Quartär – eine Standortbestimmung. Geogr. Helv. 42(2):109–115

Schoenherr AA (1992) A Natural History of California. University of California Press, Berkeley

Schubert C, Scheidegger AE (1986) Recent joints and their tectonic significance in the Coastal Range of Venezuela and in Curaçao. J. Coast. Res. 2(2):167–172

Seidl NA, Dietrich WE, Kirchner JW (1994) Longitudinal profile development into bedrock: an analysis of Hawaiian channels. J. Geol. 102:457–474

Sharma BK, Bhola AM, Scheidegger AE (2002) Neotectonic activity in the Chamba nappe of the Himachal Himalaya: Jointing control of the drainage patterns. J. Geol. Soc. India 61(2):159–169

Shields O (1997) Is plate tectonics withstanding the test of time? Ann. Geofis. Roma 40(4):955–962

Slingerland R (1981) Qualitative stability analysis of geologic systems with an example from river hydraulic geometry. Geology 9:491–493

Smith A (1986) Coarse-grained nonmarine volcaniclastic sediment: Terminology and depositional process. Bull. Geol. Soc. Am. 97:1–10

Sommerfeld A (1964) Thermodynamics and statistical mechanics. Lectures in Theoretical Physics, Vol. 5. Academic Press, New York

Spencer JW (1890) Origin of the basins of the Great Lakes of North America. Am. Geologist 7:86–97

Stanley DS (1970) The three-fathom terrace on Bermuda. Z. für Geomorphologie 14(29):196–201

Stanley DS, Swift DJP (1970) Bermuda's reef-front platform. Mar. Geol. 6:479–500

Staub R (1934) Grundzüge und Probleme alpiner Morphologie. Denkschr. Schweiz. Naturforsch. Gesellsch. 69(1):1–183

Stearns HT (1975a) American Samoa. In: Fairbridge RW (ed) Encyclopedia of World Regional Geology. Dowden, Hutchinson and Ross, Stroudsburg, Pa., 1, pp 1–2

Stearns HT (1975b) Western Samoa. In: Fairbridge RW (ed) Encyclopedia of World Regional Geology. Dowden, Hutchinson and Ross, Stroudsburg, Pa., 1, pp 666–667

Stearns HT (1975c) U.S.A.Hawai'i. In: Fairbridge RW (ed) Encyclopedia of World Regional Geology, Part I, Western Hemisphere

Stearns HT (1985) Geology of the State of Hawaii, 2nd edition. Pacific Books, Palo Alto, California

Stein G, Pflug E (1986) Höhlen im Basalt. Karst und Höhle 1984/85:235–238

Sonder RA (1938) Die Lineamenttektonik und ihre Probleme. Ecl. Geol. Helv. 31:199–238

Stevens GR (1975) New Zealand. In: Fairbridge RW (ed) Encyclopedia of World Regional Geology. Dowden, Hutchinson and Ross, Stroudsburg, Pa., 1, pp 390–400

Stillmann CJ, Furnes H, LeBas NJ, Robertson AHF, Zielonka J (1982) The geological history of Maio, Cape Verde Islands. J. Geol. Soc. London 139:347–361

Stockwell CG (1975) Canada-Canadian Shield. In: Fairbridge RW (ed) Encyclopedia of World Regional Geology, Dowden, Hutchinson and Ross, Stroudsburg, Pa., 1, pp 174–179

Storetvedt KM (1997) Our Evolving Planet. Alma Mater Forlag AS, Bergen

Storetvedt K (2003) Global Wrench Tectonics. Fagbokforlaget, Bergen

Strahler AN (1957) Quantitative analysis of watershed geomorphology. Trans. Am. Geophys. Union, 38:913–920

Sturgul JR, Scheidegger AE (1967) Tectonic stresses in the vicinity of a wall. Rock Mech. Engineer. Geol. 5:137–149

Suggate RP (ed) (1978) The Geology of New Zealand. E.C. Keating, Government Printer, Wellington, 2 vols

Taylor GI (1950) The instability of liquid surfaces when accelerated in a direction perpendicular to their planes. Proc. Roy. Soc. London, Ser. A 201:192–196

Thakur TR, Scheidegger AE (1970) Chain model of river meanders. J. Hydrol. 12:25–47

Thenius E (1974) Geologie der österreichischen Bundesländer in kurz gefassten Einzeldarstellungen – Niederösterreich. 2. erw. Aufl. Geolog. Bundes-Anst., Wien

Thom R (1972) Stabilité structurelle et morphogénèse. Benjamin, Reading Pa.

Tinkler KK (1993) Field Guide Niagara Peninsula and Niagara Gorge. McMaster University, Hamilton Ont.

Tovell WM (1979a) The Great Lakes. Royal Ontario Museum, Toronto

Tovell WM (1979b) The Niagara River. Royal Ontario Museum, Toronto

Tomkoria BN, Scheidegger AE (1967) Complete thermodynamic analogy for transport processes. Canad. J. Phys. 45:3569–3587

Truswell JF (1977) The geological evolution of South Africa. Purnell, Cape Town

Turcotte DL (1992) Fractals and Chaos in Geology and Geophysics. University Press, Cambridge

Twidale CR (1972) Structural Landforms. MIT Press, Cambridge Mass.

Udias A, Lopez-Arroyo A, Mezcua A (1976) Seismotectonic of the Azores-Alborean region. Tectonophysics 31:259–289

Wadia DN (1975) Geology of India, 4th ed. Tata McGraw-Hill, New Delhi

Walcott RI (1984) The kinematics of the plate boundary zone through New Zealand: a comparison of short and long term deformation. Geophys. J. Roy. Astron. Soc. 79:613–633

Walcott RI (1993) Extensional mechanics of continental lithosphere. Ann. Geofis. Roma 36(2):113–121

Walker GPL (1981) The Waimihia and Hatepe Plinian deposits from the rhyolitic Taupo Volcanic centre. New Zeal. J. Geol. Geophys. 24:305–324

Walker GPL (1983) Ignimbrite types and ignimbrite problems, J. Volcanol. Geotherm. Res., 17:65–88

Watson GS (1966) The statistics of orientation data. J. Geol. 74:786–797

Wegener A (1915) Die Entstehung der Kontinente und Ozeane. Vieweg, Braunschweig

Weyl R (1966) Geologie der Antillen. Borntraeger, Berlin, Nikolassee

Westercamp D, Tomblin JJ (1979) Le volcanisme recent et les eruptions historiques dans la partie centrale de l'Arc Insulaire des Petites Antilles. Bull. Bur. Rech. Geol. Min. 4(3/4):293–319

Weyl R (1975) West Indies. In: Fairbridge RW (ed) Encyclopedia of World Regional Geology. Dowden, Hutchinson and Ross, Stroudsburg, Pa., 1, pp 658–666

Wilson JT (1965) A new class of faults and their bearing on continental drift. Nature 207:343–347

Wilson JT, Russell RD, Farquhar RM (1956) Handbuch der Physik 47, Springer, Berlin, New York, p 288

Wood CA (1984) Calderas: a planetary perspective. J. Geophys. Res 89(B10):8391–8406

Wright JB, McCurry P (1970) Geology of the Zaria region. Map in "Zaria and its Region", Occ. Paper #4, Ahmadu Bello University, Zaria

Wright L (2000) Raising hot spots. Geotimes 45(11):10

Young R, McDougall I (1993) Long-term landscape evolution: Early Miocene and modern rivers in Southern New South Wales, Australia. J. Geol. 101:35–49

Zazo C, Goy JL (1994) Litoral Espanol. In: Gutierrez-Elorza M (ed) Geomorfologia de Espana. Ediciones Rueda, Madrid, pp 437–469

Zdenkovic M (1976) Entropija topografskikh karata. Informatologia Jugoslav. 9(1–4):19–38

Zeil W (1979) The Andes, a Geological Review. Borntraeger, Berlin

Zhang ZhM, Liou JG, Coleman RG (1984) An outline of the plate tectonics of China. Bull. Geol. Soc. Am. 95:295–312

Zheng FF, Scheidegger AE (2000) South Scandinavian Joints and Alpine/Atlantic-Ridge Tectonics. Ann. Geofis. Roma 43(1):153–169

Zoback, ML (1992) First- and second order patterns of stress in the lithosphere: World stress map project. J. Geophys. Res. 97(B7):11703–11728 + Map

Subject Index

Printing: Mercedes-Druck, Berlin
Binding: Stein + Lehmann, Berlin